THE GEOGRAPHER'S CRAFT

THE
GEOGRAPHER'S
CRAFT

by

T. W. FREEMAN

*Reader in Economic Geography
in the University of Manchester*

MANCHESTER UNIVERSITY PRESS
BARNES & NOBLE, NEW YORK

© 1967 T. W. Freeman

MANCHESTER UNIVERSITY PRESS
316–324 Oxford Road,
Manchester, 13, England

First published
in the United States
1967
BARNES & NOBLE, INC.
105 Fifth Avenue, New York 3

Printed in Great Britain by Butler & Tanner Ltd, Frome and London

G
70
.F7
1967

Contents

Plates

Maps and Diagrams

Preface

TEN years ago the Manchester University Press published *The Historian's Craft* by Marc Bloch, a book that by its sincerity of purpose and depth of understanding would inspire and even intimidate the author of a book with a comparable title. Marc Bloch wrote much of his work, indeed virtually all, away from sources, in hiding or even in prison. I have not been in prison, nor absent from sources, nor deprived of the comforts of modern living, nor driven by the certainty of death within a short time: almost the only resemblance between Marc Bloch's circumstances and mine has been the necessity to seek an almost monastic seclusion to read and think. I have not followed the form of Marc Bloch's book, for no two authors can work in the same way: instead I have chosen to study the work and, in so far as it seemed desirable, the lives of a few geographers chosen not for their greatness but for their variety of approach. The Press generously asked me to include some introductory and summary comments to give the book more cohesion than it might otherwise possess. Some may agree with a geographer who said at a meeting recently 'We need more geography and less about geographers': it may be so. But those who are considered here carried on their craft according to their qualities of mind and heart, and in various ways reflected the spirit of their time.

Photographs of geographers have been given in various obituary notices, or in such lives as Karl Pearson's *Francis Galton* (see p. 43). The four given in this book are perhaps of special interest and have been collected from various sources. No author ever had more helpful friends. Vidal de la Blache's portrait is taken from a pencil drawing in the possession of his family and lent to me through the kind intervention of Madame le professeur Jacqueline Beaujeu-Garnier, of the Sorbonne. It is more human in appearance than the studio portraits that accompanied the obituaries. The sketch of Cvijić appeared in *Recueil de travaux offert à M. Jovan Cvijić*, Belgrade 1924. For Sten de Geer there are some uninteresting photographs taken towards the close of his life, but the one used here was lent to me by his widow, Baroness Gea de Geer of Stockholm, following a request made by Professor Sven Dahl of Göteborg

University who responded at once to my appeal for assistance. It was taken in 1922 at Chicago. The photograph of Roxby was taken by myself on board ship, somewhere near the Bay of Biscay, in 1939. With regret, it was not possible to include any previously unpublished photograph of Ogilvie, as neither his daughter, Mrs Collins, nor his friends have anything other than the excellent photograph that has already appeared in the *Scottish Geographical Magazine* and in the volume of essays edited by R. Miller and J. W. Watson.

I am indebted to Professor E. S. Pearson and the Biometrika Trustees for permission to reproduce Figs. 1, 2 and 3. I have to thank the staffs of the Manchester University Library and the Royal Geographical Society for their help, as well as a number of friends who have discussed particular points. And I am also grateful to Mr T. L. Jones, of the Manchester University Press, for suggesting that this work should be written: to the ladies in the Geography Department who have typed it, Mrs M. J. C. Steele and Mrs A. P. Spriggs, a vote of sympathy followed by a minute's silent meditation would be a more appropriate offering.

T. W. FREEMAN

Department of Geography,
The University, Manchester,
December 1965.

Abbreviations

These abbreviations, mostly taken from the *World List of Scientific Periodicals* 1960, are used for journals.

Ann. Ass. Am. Geogr.	*Annals of the Association of American Geographers*, Minneapolis.
Annls Géogr.	*Annales de Géographie*, Paris.
Bull. Am. Geogr. Soc.	*Bulletin of the American Geographical Society*, New York.
Geogr. Annlr	*Geografiska Annaler*, Stockholm.
Geogr. J.	*Geographical Journal*, London.
Geogrl Rev.	*Geographical Review*, New York.
Geogr. Teach.	*Geographical Teacher*, Oxford (later *Geography*).
Jl R. Geogr. Soc.	*Journal of the Royal Geographical Society*, London.
Petermanns Mitt.	Petermanns, A., *Mitteilungen aus J. Perthes Geographischer Anstalt*, Gotha.
Proc. R. Geogr. Soc.	*Proceedings of the Royal Geographical Society*, London.
Revue Géogr.	*Revue de Géographie*, Paris.
Scott. Geogr. Mag.	*Scottish Geographical Magazine*, Edinburgh.
Trans Inst. Br. Geogr.	*Transactions of the Institute of British Geographers*, London.

CHAPTER I

On the Work of Geographers

MANY people have explained exactly what geography is, in works ranging from the short address to the deeply thoughtful and abundantly documented works of Hartshorne.[1] In this work it is not proposed to do that, but rather to reflect on what various geographers have done. This is not a statement of antithesis, for clearly geography in its present state has been made by those of us who write it. Nor is it claimed for one moment that those whose work is studied here were the most famous or more brilliant geographers of their times, though all had one thing in common: in varying degrees and with varying success they wrote. Some wrote too much and some less than they might have done. The geographer who is permanently absent from the printed page may be a fine teacher or a valued administrator but his light, if any, is shed only on a small circle. Of those who wrote, each had his own technique and each, though naturally influenced by the thought of the time and by the work of those who had gone before, has expressed something of himself. It may have been partly the circumstances of the time that led the Serb geographer Jovan Cvijić to turn from his much-loved physical geography to an interpretation of the human geography of the Balkans in the years preceding the 1914–18 war, but it is hard to resist the impression that he enjoyed both sides of the work. Indeed, he could not have done it so successfully unless he had thought deeply about it and—even more significant—made careful observations of the people and their ways as he journeyed year by year through the Balkans, for his work on the human aspects is too convincing and sincere for it to have been written merely as part of some deep-felt aspiration, even as great an aspiration as the emergence of a new pan-Slav state on the map of Europe.

[1] Hartshorne, R., 'The nature of geography', *Ann. Ass. Am. Geogr.*, 29, 1939, republished in 1946, 1949 and 1959; *Perspective on the Nature of Geography*, Chicago and London 1959.

I

The nature of the material

Technique there must be and to a great extent this must be observational. Under the conditions of modern education, where the success of both the teacher and the taught is judged largely by examination results, much of the learning rests on the textbook, the map and other generalized sources available to the student. And those who have to deal with senior students writing theses will discover that many of them are reluctant to believe the evidence of their own field work if it is in conflict with some report enshrined in a County Planning Office or in some textbook. In fact, everyone can make some contribution to geographical research, given the will to go out and see the landscape for himself. Piece by piece, a study of a country can be made, given infinite patience and adequate manpower; this was never better demonstrated than by the two land use surveys carried out in Britain, the first under L. D. Stamp in the 1930s and the second, with a far more elaborate scheme of classification, under Alice Coleman some thirty years later. Probably the very fact that a more intricate scheme of classification can be successfully used is itself an indication of the advance of geographical technique: in the 1930s many found difficulty in reading the map clearly but the ability to do so may now be taken for granted. But a geographer whose ability goes no further than accurate map reading would be about as elementary as a musician who merely reads the music and then thinks his work is done. Mapping the use of the land field by field gives a picture of an area that can be taken as factual. It is also dateable, for there will be many changes in the rural landscape through the years: indeed in some parts of England, notably the southeast and the Midlands, people are beginning to wonder how much of the rural landscape will be left by the end of this century.

Field work is an activity to which virtually everyone pays lip service but fewer people actually practise. It is endlessly varied in scope. Famed travellers of the eighteenth century went through England and reported on their journeys to delighted readers whose range of travel was necessarily limited. In the nineteenth century, travellers went further and Sir Francis Galton, whose adventures are discussed in chapter 2, was only one of many intrepid souls of his day. As the seas were opened more and more to shipping, and traders, missionaries and colonists penetrated the greater part of the world, so knowledge of the whole world came within the grasp of people able to study maps or eager to read the adventures of the explorers. A world view became fashionable and the cultivated man of affairs was photographed or painted with a globe in the background. What geography owes to the explorers cannot be estimated: it was they who brought the observations that were pieced together by the compilers of atlases and writers of books: true, the atlases were highly generalized from the paucity of observations into a deceptive simplicity. As more was known of climate, vegetation and soils, so the idea of relationship grew, especially when the Berghaus atlas was published in 1837,[1] and from 1859, when Darwin's idea of the evolution of species and, still more, of the mutual interdependence of all life, was published. Observations brought by the explorers appeared to give confirmatory evidence. Though a question that had engaged the minds of thinking men from the beginning of time, the question of man and his environment became of greater interest than ever before. What in fact were the resources of the world? Malthus looked forward with apprehension to a time when the number of people in the world would be too great for the resources it possessed and the 'population problem' became a subject of frequent discussion much as it does at the present, though the world population is now probably three times as large as it was a century ago. Our knowledge of the world, built up by one observation after another, in short by field work, seems great, especially if one goes to libraries and sees rows of volumes of geographical works on various countries.

In fact, is it so great? Arrogance not infrequently conceals vast stores of ignorance and generalizations are best made when the data are inadequate. The author remembers that in his first year at the university C. B. Fawcett gave him a shock by explaining that the smooth lines on the rainfall map of Australia were indicative not of the average rainfall but of the lack of knowledge of it. So little is known of much of the world and so many prognostications prove to be false. At the time of the Czarist régime, Russia seemed an exhausted and overpopulated land, unlikely to make much further progress, riddled with poverty and—partly for this reason—ripe for revolution. In fact its resources were largely untapped and underdeveloped when its population was half what it is now, and it is obviously for a purpose that its government sends geographers out

[1] See p. 59.

as field workers in research teams to developing areas. Those who wish to learn more of Russia can at least read some of the abundant geographical literature now available, much of it translated. But there are other areas of which little is known; and of the larger countries of the world, none seems more enigmatic than Brazil, whose agricultural possibilities would appear to be at least considerable. One cannot assume that the food producing capacity of the world has been reached, nor that the resources of minerals and power are within sight of exhaustion. And what of India? True, many geographers have worked on it and given interesting material but—as A. G. Ogilvie pointed out[1]—the sum of knowledge of this vast sub-continental area is woefully inadequate. That India has a population problem is undoubted, yet more regional surveys are needed for the planning of its future progress.

For much of the world geographical information consists of half-truths which conceal as much as they reveal. On the other hand, much devoted survey has been done, not least in India by devoted workers under a British *raj* (which did not, as some people allege, spend its entire time in social and athletic pursuits), and by the authors of the Census reports.[2] In India, as elsewhere, much of the challenge lies in the rapidity of change, a subject to which one must frequently return. All through the sub-continent, the geographers of the Indian and Pakistani universities have no fear of a lack of subjects needing research.

A first necessity is to ask the right questions and to seek the answers that it may not always be possible to find. When writing the regional section of the Admiralty Handbook on China during the 1939–45 war, one longed for some local accounts of the scene to supplement the limited amount of information one could glean from maps and photographs: searches in the libraries revealed a few pieces of local field work, mainly by G. B. Cressey and Chinese authors. Analysis of every possible photograph gave some impression of the character of the coasts, especially between the Yangtze and the Si rivers and interviews with those who had sailed the coast gave more. Yet when all this was acquired, how certain could one be that the

---

[1] See p. 182.

[2] A number of excellent textbooks are available, notably Spate, O. H. K., *India and Pakistan: a General and Regional Geography*, London 1954, 1957; Farmer, B. H., *Pioneer Peasant Colonization in Ceylon*, Oxford 1957.

inevitably generalized descriptions bore more than a superficial resemblance to the real landscape? And a similar difficulty arose when
the statistical material was studied: so often to geographical writers
these seem to offer some kind of certainty as basic data. For China,
virtually all the estimates of population available in the 1940s came
to approximately 400 millions, but when they were examined the
figures for various provinces proved to differ so widely from one
source to another that their addition to one standard 400 million for
the whole seemed almost a miracle. And when the populations of
individual towns were investigated, even wider disparities appeared
with estimates ranging from 150,000 to 625,000 for one town on the
southeast coast of China. It may be that the former estimate dealt
with the people living in a smaller area than the second and larger
estimate, for in China as elsewhere in the world many places spread
along coasts and river valleys and as we are convincingly informed by
descriptions and shown by photographs, thousands live permanently
on boats. Even so, certainty does not exist in such areas, for varying
reasons: if the more recent estimates of the population of China, at
over 600 millions, are to be believed those of 400 millions must have
been wildly wrong. It has been said with some justice that estimation
of the population of China resembled a parlour game unencumbered
by any definitely established rules.

In varying degrees there are similar difficulties elsewhere, as people
may wish either to avoid a census, if it seems to be a prelude to taxation or military service, or to be counted twice if it foreshadows
voting rights or financial benefits: and other difficulties of enumera-
tion arise in the case of migratory populations, as any District Officer
in the former British colonial territories would explain. Nor need one
go to places so remote for examples of inaccurate statistics, for even
in Britain the area of land used for various purposes was imperfectly
known until recently, as R. H. Best has shown.[1] Not all the reasons
for inaccuracy imply any malign purpose on the part of farmers
though it is interesting to note that when feedstuffs were rationed
during the 1939–45 war the area of land suddenly increased. In
Britain so much controversy centres around the present and future
use of land that the initial need, as Best has said, is 'to get our factual
house in order'.

[1] Notably in Best, R. H. and Coppock, J. T., *The Changing Use of Land in Britain*, London 1962.

The wish for certainty

Certainty is as elusive as the white bird of truth. Marc Bloch[1] gives
an example of an execution during the September massacres of 1792
in France, when the head of the Princess de Lamballe was paraded on
the end of a pike under the windows of the royal family. Is this true
or false? An historian dealing with the period does not venture an
opinion: had he been there to see the ghastly event, and had he
recorded it with scholarly detachment at the time, the evidence for
future historians would have been valid and certain. In short, the
historian present at the actual event would have been the perfect field
worker, observing and recording what he saw. It is not easy to induce
students to note everything down as they go along, and one admires
the pertinacity of those who, like a group from Manchester University
in 1965, brought back completed observations on maps and farm
interview forms spattered by the rain for which the west of Ireland
is so justly renowned. Oddly enough, a fine enterprise in field work
was carried out in the west of Ireland during the 1890s, in the effort
to find out the economic and social conditions of an area subject to
chronic poverty and occasional near-famines. Under the general
management of the Congested Districts Board, investigators went
with a standard questionnaire to every area in which the assessed
value of the land and houses was below 30s a head, and enquired into
the size, crops and stock of all the farms, the sources of supplement-
ary income such as local labouring, fishing, seasonal migration, home
industries, emigrants' remittances of the population and the incidence of extreme poverty. An estimate was made for each area of the
normal income and expenditure. No doubt some of the investigators
were more observant and careful than others, though as the present
writer read the Reports all save a few hastily compiled at a late stage
appeared to be as accurate as one could reasonably expect; naturally
enough, the observation on which least reliance might be placed is
that on poverty, for so many imponderable factors influence it, such
as the mental and physical health of the people studied, their tendency to save or to spend, their acceptance of low living standards.
If as is commonly believed many residents in suburbs are driven by
an urge to appear wealthier than they in fact are, in a peasant community some may be influenced by a social ethos to appear poorer

[1] Bloch, M., *The Historian's Craft*, Manchester 1962, 51.

than they in fact are. Field work in the west of Ireland is made easier by the hospitable and charming qualities of its people, as many parties from various universities have found in recent times. This example from the west of Ireland is not given as something unique for there were many such enquiries in the nineteenth century and some of later date, such as the reports of a similar Congested Districts Board for Scotland from 1911, some of which could provide excellent material—so far almost untapped—for historical geographers.

In short, the basic need for geographical study is field work. The nineteenth century desire to comprehend the entire world is natural, even laudable, but there has been a growing reaction against the superficiality of much that is taught on a vast world canvas. It is part of the work of a university teacher to show his students that everything must be questioned, for little can be done with those teachers and students who are authoritarian in mental outlook. Even more trying is the university teacher who spends his time in demolishing imaginary opponents. But when one begins to study some quite elementary point by re-assessing the evidence, unexpected results may appear. For a long time, authors wrote—probably quite thoughtlessly—that high densities of population in rural areas were indicative of fertility of the land and intensity of farming. It may well be so but the statement needs some qualification. In studying the case of Ireland, away from the cities and excluding the few and comparatively small rural areas in which people working in occupations other than farming live, it was found that the range of population density was from less than 50 to more than 400 per square mile. Even since the Famine period of 1845–51, the rural population of Ireland has declined, and clearly no uniformity has been achieved. The answer is quite simple: the density is related to the size of farm held and the type of farming practised. In parts of the west, the average size of farm may be 7–8 acres, and therefore a square mile may have within its 640 acres as many as eighty farms: if, as is usual, there are four to five persons in each household, a density of 400 to the square mile may result. At the other end of the scale, in areas such as the grazing belt in the counties of Meath, Dublin and Kildare the farms may cover as much as 150–200 acres so that within a square mile there may be only three to four farms. Grazing farming is economical of labour, and probably each farm will have only one or two paid labourers, especially if family help is available, so the population density may be only about fifty per square mile. In some parts of the west the soils are developed by placing seaweed and sand on ill-drained clays or patches of peaty humus between solid rocks or large stones removed by the patient labour of generations of farmers into large walls. Mechanization of any description may be impossible, and each farmer can only cultivate as much as he can cover annually with a spade. Many, in fact most, will have some access to rough grazing but if one excludes these in assessing the density per square mile of occupied land, remarkably high densities result. And there are comparable conditions in the crofting areas of the Highlands and Islands of Scotland. Geographical study of such areas reveals a relationship of the rural life to inherent physical conditions, a response that has been worked out through the years to the particular qualities of the environment.

Further questions immediately suggest themselves. As the farms are so small, poverty must inevitably result unless some extra resources can be found within the local area, or by migration of a seasonal nature. In fact, there is a wide range of extra resources, such as working intermittently on making roads, seasonal migration, fishing, domestic industries such as weaving, knitting or embroidery, peat gathering or occasional labouring for the richer people in the community. Money comes into such areas from emigrants who have prospered in Britain or the United States, and therefore the community is only partially supported by the land, for in addition to all these varied sources of extra income there are various governmental subsidies and benefactions. At various times the solution of taking the large farms and demesnes of the landed gentry and dividing them into new farms for the people of 'congested' areas has been suggested, and this seemed to be one possible means of alleviating poverty to the workers of the Congested Districts Board. To some extent this has been done and former estates, of which in some cases little use was made, have been divided into geometrical fields, some of them in strips from the valley floor to the moorland, not always with fortunate results as in some cases it may be as much as a mile from one end of the farm to the other. There have also been experiments in settling people from the west in the richer eastern areas, by dividing up the large grazing farms—contemptuously spoken of by some as 'ranches' though they bear little resemblance to a ranch of the New World type—into compact farms of some 30–40 acres.

As one surveys the rural scene, nothing is fixed, nothing is permanent. The author well remembers his surprise on a first traverse of Ireland with a brief sojourn in the west at the number of new houses built all over the farmed landscape and later work quickly showed that the field pattern, in fact the whole distribution of the rural population, had changed markedly and recently. And when the study was extended to villages and towns, there too the changes were considerable, as scores of poor squatters' hovels in the notorious 'cabin suburbs' on the fringes of settlements had been abandoned and at least partially removed—unfortunately not completely as land in Ireland is not as scarce or as expensive as it is in Britain. No textbooks had told him all this: in fact they had told him some strange things about Ireland. As one reads works of the 1930s, the view is inherent that the loss of population from rural areas was everywhere to be regretted[1] and at a time in human history when vast population increases are in progress and expected to continue, it is also clear that agricultural work will not provide more jobs than it does now, except in a few areas where market gardening may replace less intensive forms of farming. Generally, the number of people employed in agriculture will decline, partly through mechanization and partly through the withdrawal of people from the poorest areas. Nevertheless, in some rural areas efforts are being made to arrest the decline: in a parish of southwest Donegal, remote almost beyond most people's imagination, vegetable growing has been made the basis of food processing industry and home weavers of tweed have been drawn into a factory where designs and standards of workmanship can be guarded and the products—of excellent quality and great appeal—can be sold co-operatively. There are further schemes of co-operative land improvement and sharing of produce. Initial difficulties such as remoteness, an excessively rainy climate, large areas of poor soil, the loss year by year of many of the more enterprising people by emigration, all mean little to a spirited few anxious to prevent the gradual rotting of a community and willing to use the few advantages the area possesses, actually or potentially, which include the existence of patches of good soil, on which vegetables can be grown, and the survival of a home craft that gives tweeds saleable in the international market and a reasonable wage to those who practise it. And this raises the whole question of whether similar enterprise might not be possible elsewhere. In a modern society

[1] In Sweden and elsewhere, many farms are disappearing.

possessing great resources, with high levels of taxation, it is easy to pour doles into poor areas, but doles cannot provide a permanent solution for such problems as endemic poverty.

Nevertheless, there are many areas where sharp declines of population are likely, and probably unavoidable. One by one islands off the west coast of Ireland are evacuated, though sheep may still be left on them to crop the herbage. Similarly many alpine areas such as the one mentioned on p. 178 will be occupied only as summer pastures instead of as areas farmed throughout the year. In Norway many of the summer pastures are no longer used, for the farmers who own them can buy the extra fodder needed and spend their time more profitably on some form of activity such as apple growing, tomato production in glasshouses or mink rearing for furs. In France, with its vast range of agricultural soils, some poor areas now have a population no longer regenerating itself, consisting largely of widows left from two wars with a few old couples. When a farm falls vacant, no purchaser may be found, and there, as in some parts of Ireland, it may be added to another holding and in many instances allowed to decay. All who are familiar with marginal lands through mountain walking will know of farms that for one reason or another have been abandoned because no one could be found to occupy them though, for a time at least, the land may be used and its fertility retained. Remoteness, the lack of roads, distance from schools, shops and services, all may be factors more crucial than quality of land in determining rural population decline. And inevitably the whole conception of a reasonable standard of living changes substantially from one period to another. What was regarded as adequate in the days of the Congested Districts Board is far below what would be regarded as adequate now, for the luxuries of yesterday have become the necessities of today, and the luxuries of today are the necessities of tomorrow. Nor can there be a difference between standards of living among rural and urban populations for piped water, electricity, social and athletic facilities, entertainment, education and much more are as essential to the rural as to the urban population: no enemy is feared as much as boredom. Only gradually has the appreciation of such needs come to the countryside but in time it must come everywhere through the press, radio and television, as well as through education. Along with this, government services are expected to cover a steadily widening range of opportunities and to alleviate an increasing number of the vicissitudes of human life.

Environmental influences

It may seem as though the influence of the environment on man steadily diminishes as life becomes more sophisticated and less elemental, in short as more people insulate themselves from natural hazards by protecting themselves and their animals from cold or drought by heating and piped water. Study of primitive societies has often been advocated as a means of appreciating the relation of people to their environments, especially as most such societies draw little from the outside world. The argument between determinism and possibilism has been so endless that it is unnecessary to raise it here: in an extreme form, the determinist case is given by Ellsworth Huntington,[1] whose work stands out in contrast to that of others such as Vidal de la Blache or Cvijić. If one considers some of the most advanced agricultural societies of the world, it is apparent that their practices are geared to a finely discerned understanding of economic possibilities: the dairy farmers of Denmark, the glasshouse cultivators of south Holland or even the flower growers of Cornwall and the Scilly Isles can only exist profitably if good communications are available to transport their produce to the townspeople who consume it, but they have used natural climatic and soil conditions as a basis for such activity. And that is not all: in fact they have fertilized and developed their land, drained it, irrigated it, even heated it artificially in greenhouses and protected it from pest infection by modern scientific means. Vast areas of land that now seem rich, such as large parts of Denmark, owe their fertility to human skill in making soils, though even the Danes have left their poorest glacial sands for use as forestry plantations. In the nineteenth century flush of scientific enthusiasm many thought that the earth made man, though in fact man made the earth, subject to certain limiting factors. Possibly no landscape seems more thoroughly man-made than that of south Holland, where the neatly defined edge of the Hague lies beyond an intensively-cultivated belt of market gardens, each having its outdoor patch and its glasshouse heated by coal with a consequent forest of chimneys. Patently adverse physical circumstances have been overcome by the control of water in its passage through narrow, and now rarely used, channels between the holdings to the broader canals and finally to the major outlets such as the Rhine deltaic channels, long since controlled

[1] See chapter 5.

into waterways that in fact are canals. Dutch towns can only be expanded with careful planning based on water engineering and the dunes on the coast may be water gathering grounds as well as natural ramparts against the sea, so access to the beaches is permitted only at certain points. The triumph over natural hazards seems complete though disasters such as the flooding of 1953 are a reminder that occasional hazards such as a combination of an unusually high tide with gale force winds can be catastrophic.

Deviations from normal or average conditions can be a subject of some fascination as well as great human significance. Climatic study has been made possible for vast numbers of people by striking averages so that the essential characteristics of particular climates can be readily understood. But as more statistical information has accumulated averages have carried less and less conviction: variations in such matters as the dates of the first and last frosts, the depth of snow from one year to another, the rainfall experienced in the heaviest known falls, the longest drought periods, all may be of great human significance. The *Atlas of Finland* shows among many fascinating maps a series of the Baltic in average, severe and mild winters, and if one travels along the Norwegian coast on more than one occasion as the author did in two Aprils, the same areas will have different snow conditions according to the weather of the early months of the year. Workers on glaciers have now convincingly shown that those of the northern hemisphere are decaying, but whether this will continue indefinitely only time will show: in 1965, the quantity of Arctic ice was said to be increasing. But those who through the years have patiently watched such processes have added considerably to the knowledge of glaciation: in fact few physical processes can be observed in action over a relatively short period such as twenty or thirty years and probably for this reason some efforts have been made to reproduce them experimentally,[1] though with only moderate success.

Having accumulated data, the need is to give it form and cohesion. The records of nineteenth century explorers, the material acquired from the various meteorological stations throughout the world, the statistical and other enquiries of various governments, the observations of individuals in many areas, all provided the basis for an initial regionalization of the world that is familiar to those who have studied geography at all seriously. But once an initial synthesis is made, its

[1] Cf. p. 173.

limitations become apparent, so much so that many begin to wonder if there is any possibility of making a synthesis at all. When this happens, some geographers will become systematists, taking one aspect of the subject as their specialism and regarding their predecessors as obsolete. This development will undoubtedly be fruitful and already one could instance many examples of progress, say in the work on urban geography gradually developed since R. Blanchard wrote his work on Grenoble[1] in 1911. Many studies have followed that on Grenoble in taking an historical view and observing how the town has grown from one age to another, but other methods of investigation have also been tried with considerable success. These include the careful, statistical study of town centres to discern the relative attraction of various streets for shopping and service trades, the study of suburban shopping centres and much more that is not only of some interest in itself but also as a basis for planning. Historical studies have shown the steady outflow of population for various reasons, including the change of land use in central areas as houses were bought for demolition and their sites used for new roads, commercial premises and industrial purposes. Every town study shows the ephemeral nature of the existing land use as so much has been altered within a relatively short time: on the one hand this inspires the reflection that much of value has been lost and, on the other, it brings forth the hope that drastic remodelling may be possible. On a first visit to London, the author saw the last of the famous Regent Street crescent and its replacement by the uninspired pseudo-classical buildings of the mid-1920s: should this have happened? None could regret, however, the bulldozing of the nineteenth century slums of the industrial collar which surrounds the central business district of Manchester and its replacement by blocks of flats, buildings of a much-enlarged university and new ring roads. And the changes in Birmingham have been at least as great, if not greater. What has been seen already is only a small part of what is to come, because most towns have schemes of regeneration, and beyond the areas to disappear under slum clearance schemes there are others of drabness and incipient decay whose life must be limited.

Earlier it was noted[2] that the earth was made by man—at least with some limitations. If that is true of the rural areas, it would appear to be even more certain of the towns, and yet closer investigation shows

[1] Blanchard, R., *Grenoble*, Paris 1911.   [2] See p. 11.

that physical features can be a marked influence on the form and shape of towns. Bridging points with dry ground on each side, a defensible plug of rock on which a castle could be placed or a sheltered haven for shipping, all these and many more proved attractive to the builders of towns. And some features of the physical geography may act adversely on certain parts of towns: until recently the areas near the Thames in Oxford were not favoured for housing owing to the danger of floods while immediately beside them the late-nineteenth century suburbs of north Oxford grew on a gravel terrace only a few feet higher in altitude. In many towns the higher ground is favoured for suburbs and as one ascends in altitude, so one ascends in the social and financial scale. In the government report on the northwest of England, published in July 1965, it is reported that the death rate is higher than in many other parts of the country but no one knows exactly why. Among reasons which will be sought one is surely the incidence of smoke which may lie for days over industrialized Pennine valleys or be distributed by winds eastwards towards such towns as Oldham and Burnley on the fringes of the Pennines: to remove misapprehensions, one must add that this may be only one of many reasons, which may include the relatively high proportion of old people in a population long subject to heavy outward movement of younger people. The actual layout of towns may show a clear relationship to physical features: in Glasgow some of the drumlins which form part of the site are laid out attractively with curving roads leading to stately crescents around gardens on the summits while others are seared with a grid-iron pattern of streets, some with extremely steep gradients. And American cities have a grid-iron pattern almost everywhere: so for that matter has the countryside. In south Manchester, however, the land between the Medlock and the Mersey is so flat that there was nothing to prevent the spread of a grid-iron pattern of roads, which has developed to a considerable extent though fortunately some roads preserve the deviations they originally acquired as country lanes. All who live in large cities now see them changing markedly, but the new blocks of flats or houses with gardens, if they are to form a pleasant and harmonious town environment, must be so placed that they draw out the interesting features of the physical environment. In Holland, the flatness of the landscape is used effectively as a setting for many a village clustered around a church with a noble spire.

The field observer

For many students in universities and training colleges, the first (and probably the last) piece of research done is a dissertation for a degree or diploma. From the author's experience even students of quite moderate ability can with patient field work and enquiry present interesting accounts of their own town or countryside. But as the years have passed, the area covered has steadily diminished from eighty square miles or so to an individual town or a few square miles of countryside, at least in some universities. Breadth versus depth of research, or at least enquiry, may open arguments of inestimable futility but those who work on a small area, or a single town, even one quite small in size, will probably find it offers more problems and greater interest than might have been expected. And this is particularly true if one takes an historical approach. The analysis of the ancient small town of Alnwick by R. G. Conzen[1] shows the richness of historical circumstances that have made the town an interesting but varied whole: in Conzen's monograph the historical material is used geographically, with the emphasis on the distribution of houses, public buildings, roads, gardens, open spaces and other features that make up a town. Obviously this is not the only way of studying a town: what, for example, could be revealed by a detailed sociological or economic survey of such a place? But geographers do themselves no good by launching themselves into sociology—that is, unless they intend to become sociologists—as a kind of sideline, for the sociologist's contribution is distinct from that of the geographer, though related to it.

If one considers problems of industrial location, similar problems appear. A factory may exist in isolation, though more probably it will be in some kind of association with others, possibly because power or some raw material is locally available or can be brought from elsewhere by sea, canal, road or rail. Even if the initial advantages of location disappear, either because the raw material is exhausted or because sources of power change and become available widely instead of only in certain places, the factory may still retain its original location. But there may be catastrophic changes, of which not the least obvious is the fall of the cotton industry of Lancashire and northeast Cheshire to a fraction of its former strength, but without

[1] 'Alnwick, Northumberland: a study in town-plan analysis', *Trans. Inst. Br. Geogr.*, 27, 1960.

the disastrous economic results that might have been expected as many new or expanding industries have been settled in old cotton mills, substantial in size and cheap in market value: this type of inducement will be strong in periods when factory building is restricted by government policy, shortage of sites or other controls. Industrial diversification is desired almost everywhere to avoid evils such as sudden decline or even slow decay due to the exhaustion of resources such as coal, and at one time some geographers regarded a range of industry corresponding to that of the country as 'balanced'. An area such as South Wales obviously depended to a dangerous extent on coal mining and needed more light engineering, clothing, food and general consumer industries to bring greater stability: a 'location quotient' showing its distribution of workers between mining industries and services would show the dangerous local dependence on coal mining. This relatively simple type of analysis has some value, though some diversity between one industrial region and another may be economically desirable in view of the natural advantages each possesses for certain forms of production: nevertheless for a variety of reasons the industrial diversification seen in many areas of Britain within a comparatively short space of time is remarkable. And this strengthens the need to investigate particular factories, to find out why they are in their present location. The reason for the existence of some industries may be apparently trivial, even extending to the choice of a pleasant area to live in. But as S. H. Beaver has shown,[1] for some industries the locational possibilities are so markedly restricted that geographical environment is a permanently significant influence. Aluminium smelting requires great quantities of heat provided by low voltage direct current, so it is carried out beside generating stations as this type of current cannot be readily transported. Oil tankers are now so large that they can only berth at a few points and oil refineries must be on coastal sites or linked to the berth by pipeline. But for industries less obviously restricted, new means of communication may alter the position of a town considerably: for example the completion of the motorway through central Lancashire in the 1960s brought hopes of industrial revitalization to areas in the central Lancashire coalfield, damaged by the decline both of mining and cotton. More and more the control of changes is regarded as a responsibility, in part at least, of governments though a vocal if

[1] 'Technology and geography', *Advancement of Science*, 17, 1961, 314-27.

diminishing few would argue that such matters are best allowed to take their natural course.

'Geography is the science of locations. Regional geography classifies locations and theoretical geography predicts them.' So says Bunge[1] in a book which shows some of the possible uses of mathematics in geography and discusses the visual presentation of distributions in maps and diagrams. This subject is too vast to be treated here, though the underlying need to assess relationships in terms of distance and possible movement of people and commodities has long been recognized. If, for example, it emerges that country towns are a certain distance apart over a large area, that cannot be fortuitous: if, for example, it appears that in the west of Ireland areas that can be called towns are few and distant, but that between them places that are villages in size have facilities that are normally associated with towns, that again cannot be fortuitous. Obviously some factor must be responsible for the smallness of the market centres, the central places in the west of Ireland, and what else can it be but the lower buying power of the population? Time distance may mean more than actual distance: a commuter into central London may choose a more remote location, and one involving more travel cost, in preference to a nearer one because it is easier in access.[2] The range of shopping facilities available in a village will be influenced by the number of customers drawn into it by the various means of transport and the adequacy of the roads available to the people of the surrounding countryside. All this, and much more, can be assessed mathematically, not only from observations of what has emerged already through 'trial and error learning' to quote one catchword of educationalists, but through reasoned prognostication of future growth at a time of marked population expansion and the associated inevitable spread of urban society. Without such foresight, many new developments will not provide reasonable living conditions for the people drawn into them, as experience of some of the new housing areas of the 1920s in Britain convincingly demonstrated: there must be work available either locally or in places sufficiently close to be reached rapidly and not too expensively, there must be shops and a wide range of social, athletic and cultural pursuits if people are not

---

[1] Bunge, W., *Theoretical Geography*, Lund Series in Geography, Ser. C, General and Mathematical Geography, No. 1, 1962, 195.

[2] In the remoter places houses may be cheaper.

to die of boredom and the youth destined to fall into delinquency. What size of a town is likely to provide such facilities? In Britain, the suitable size for a new town or an expanded town appears to have risen sharply from 50,000 in the late 1940s to 80,000 later and now to 150,000 or even 250,000 in some cases. Quantitative measurement has its obvious application to future developments, but it is also of use in historical geography: for example, what proportion of the population lived in towns in pre-industrialized societies? From the author's investigation of Ireland at the beginning of the 1840s, when factory industry was meagre, the town population was less than one-fifth of the whole and there was ample evidence that the towns were a refuge for the destitute from an overcrowded countryside.

The past and the future

Claims to predict the future on a mathematical basis can only be tested by the lapse of time, though as always adequate knowledge of what exists at the present is a guide for future action. Where are the problem areas, where are the areas of affluence and growth? Patrick Geddes, a pioneer of planning, spoke constantly of 'survey before action', and among many planners the realization of the present is that such planning cannot for all time be tied to the local administrative units but must deal with larger regional entities such as the northwest, the northeast, the southeast with London, and others of similar scope in England. The word 'region' may be used in many contexts with many meanings, but it still has meaning and relevance to the lives of the people. It may develop a kind of specialized 'regional science' of its own, at once historical, economic and geographical: such trends are already apparent in America. And the idea of 'environment', basic to geographical study in the days of some modern pioneers, remains compelling, for in some universities efforts to give a broad and liberal education are focused on 'environmental studies', which may include almost anything that affects human life, history, geography, economics, local government, law and much more. How much a student knows—really knows—after taking such a course in 'environmental studies' is open to question: no geographer of sense has ever claimed that he alone can interpret environment. Division between subjects is artificial in any case and much of interest can come from those whose interests are marginal or dual. On this Marc Bloch has an interesting comment:

... each science, taken separately, finds its most successful craftsman among the refugees from neighbouring areas. Pasteur, who renovated biology, was not a biologist—and during his lifetime he was often made to feel it; just as Durkheim, and Vidal de la Blache, the first a philosopher, the second a geographer, were neither of them ranked among the licensed historians, yet they left an incomparably deeper mark upon historical studies at the beginning of the twentieth century than any specialists.[1]

This statement was found in an incomplete form on a loose sheet of paper when Bloch's book was prepared, and may not be in a finished form: the comment on Vidal de la Blache may well be exaggerated, as he was hardly a refugee from geography, but rather one who had a combined interest in history and geography, expressed so remarkably in his Atlas[2] and in some of his other works.

Dual interest is inherent in all study of historical geography. The evolutionary view of landscapes, as gradually developed and changed from one age to another, implies that it should be possible to reconstruct the landscape of any past period from documentary sources, maps and possibly excavation also. The main features of the geography of the Roman period in Britain are known and mapped, and several of their town sites and roads survive to this day: on the Roman wall in Cumberland and Northumberland it is easy to imagine the Legions guarding the south against the north. Work has been done on a comprehensive scale on the Domesday period under the general editorship of H. C. Darby[3] and some fine work has been done on medieval villages by M. Beresford.[4] It may be true, as some historians suggest, that the Black Death of 1349 saved England from famine, as the food resources of the country were taxed to the limit, but Beresford has shown that not all the villages were abandoned as a result of the Black Death. In spite of such changes, the main theme in British historical geography is continuity, for many sites have been occupied for 1,000 years and more, and some churches are on sites

[1] Bloch, M., op. cit., 21–2 (footnote).

[2] See p. 52.

[3] Darby, H. C., *Domesday Geography: Eastern England*, Cambridge 1952, 1957; *Midland England*, ed. Darby, H. C. and Terrett, I. B., Cambridge 1954; *South-east England*, ed. Darby, H. C. and Campbell, E. M. J., Cambridge 1962; *Northern England*, ed. Darby, H. C. and Maxwell, I. S., Cambridge 1962.

[4] Beresford, M., *History on the Ground*, London 1957; *Lost Villages of England*, 1954; and with St. Joseph, J. K., *Medieval England: an Aerial Survey*, Cambridge 1958.

that may have been Christian since Saxon times, even if all traces of buildings dating from such remote ages have long since disappeared. There are vast problems awaiting research in historical geography, not least of the nineteenth century for which much useful contemporary material in the form of maps, statistics, government reports and—now fast disappearing—houses and factories on the ground exist. Modern techniques of mapping can be easily applied to nineteenth century statistical material, and a great deal of contemporary map material lies unstudied in libraries, not least in the Map Room of the British Museum. But anyone wishing to practise the craft of the historical geographer must follow an historical technique as well as his own technique and without malice one could add that some historians might use geographical material more intelligently, and not merely as an occasional *deus ex machina*, to explain the apparently inexplicable. And it is strange that some historians producing fine and scholarly works illustrate them by maps that appear to be rejects from the lesser newspaper offices.

Historical geography offers the attraction of a retreat from the present as it asks the question, 'What was the landscape like then?' as the historian asks 'What happened and why?' But in either case study of the past illumines the present and as the geographer looks over a landscape or visits a town, it is rewarding to reconstruct in imagination its evolution: if one visits Caernarvon or Conway in North Wales, the castles and the walls show the clear pattern of a medieval town beside a navigable river whose crossing was carefully controlled from the turrets and walls. In many such places there are still relics of ancient street patterns, and some main roads still preserve the line of Roman routes: beside the Roman wall there is a military road of the eighteenth century. The older settled lands of the world, not least in Europe and especially the Mediterranean lands, have age-old features in their landscapes, and this is not less true in Asian countries such as China, where some lowlands have so intricate a series of man-made waterways that it is impossible to reconstruct the original drainage pattern at all. Those who develop an historical approach to landscape can also see changes before their very eyes, especially in the large towns and cities, but many are sadly unobservant and appear to have little interest in indications of the social revolution now proceeding. The present author has less faith than many others in the prospects of foreseeing the future, even with the interesting mathematical techniques devised by Bunge and many more,[1] for the outlook of a new generation on planning may be entirely different from that of the present time. But if prophecies of a higher standard of living and greater leisure for the developed countries of the world prove accurate, there must be great changes both in towns and in the countryside. Beyond all such possibilities, the problems of the underdeveloped world remain and the eventual removal of poverty and malnutrition from the world will only be achieved with many years of unremitting effort. The natural resources of the world are not fully exploited as yet and the fears of some writers that the population is growing beyond the resources may prove as groundless as those of Malthus a century ago. Nevertheless, nobody with imagination can accept with equanimity a state of world society in which a substantial proportion of the population live in a state of chronic poverty and undernourishment.

The geographers considered in this book came from varied backgrounds and met the challenge of their times in varied ways. Galton shows the Victorian zeal for travel and experiment; Vidal de la Blache looked from his well-loved France outwards to the whole world, only to return again to the homeland; Cvijić based immensely important work in political geography on his own carefully accumulated observations; Huntington showed the courage of an adventurous mind in his vast synthesis of human life; Sten de Geer painted on a large canvas in some of his work but left a more abiding impression through his fieldwork in Sweden; Roxby and Ogilvie gave stimulus to British geography at a crucial stage in its modern development. All had a sense of purpose with a wish to serve in their day and generation and have been chosen here as representative rather than dominating figures of their time.

[1] See p. 17.

c

CHAPTER 2

Francis Galton

A VICTORIAN GEOGRAPHER

IT was not as a geographer that Francis Galton acquired lasting fame: it was as a geographer that he acquired his Fellowship of the Royal Society in 1860.[1] Almost all his geographical work was published between 1852, when he wrote an account of his expedition into the interior of southwest Africa, and 1863 when his meteorological atlas appeared. 'Hereditary talent and character' was the subject of a paper in *Macmillan's Magazine* in 1865, and four years later his book on hereditary genius appeared, followed in 1874 by *English men of science, their nature and nurture*, in which he analysed the intellectual and physical qualities of his own and many other scientific families of the time. Among these were the Darwins, descended from Dr. Erasmus Darwin, F.R.S. (1731–1802), a physician and poet who was a friend of Joseph Priestley and James Watt: a son through his first marriage was the father of Charles Darwin (1809–82) and a daughter of his second marriage, F. A. Violetta Darwin (1783–1874) married Samuel Tertius Galton (1783–1844), a successful banker whose wealth was due in part to the sale of muskets in the Napoleonic War. In 1822, when Francis was born, his parents lived at 'The Larches', a house one mile from Birmingham on the Warwick road, then surrounded by fields and woods but now submerged by streets and factories. In 1824, S. T. Galton bought a second residence, Claverdon, a country house near Warwick; in 1831 he gave up banking and in the following year he went to live in Leamington, then a small but growing place. Francis Galton, the youngest of his

[1] And not in 1856, as stated in various places. The certificate of a candidate for election gives his title as 'Assistant Secretary, Royal Geographical Society' and his qualifications are listed as 'Recipient of the Gold Medal of the Royal Geographical Society, for having explored at his own cost the central part of South Africa; Author of the "Narrative of an Explorer in South Africa", "Art of Travel" etc. etc. Distinguished for his acquaintance with the science of Geography. Eminent as an African explorer and geographer.' The supporting signatures include that of Charles Darwin. He was elected in June 1860. For this information thanks are due to Mr J. C. Kenna, who consulted the archives of the Royal Society.

nine children of whom two died in infancy, was born into an atmosphere of wealth and culture on 16 February 1822 and died on 17
January 1911 at Haslemere, having for much of his life 'enjoyed ill-
health', to use a contemporary saying: happily immune from the
burden of earning his daily bread, he worked extremely hard for a
number of societies and pursued many research enquiries avidly,
alternating such activity with protracted holidays. He could be taken
as a supreme example of the Victorian man of letters, leisured yet
active, nervous but purposeful, versatile yet thorough in his work.

Precocious in early childhood, Francis Galton went to school
early for companionship and occupation, and in his reminiscences
spoke of the first few happy years as spent in 'what was virtually the
country', that is, mainly in what is now the Sparkbrook area of
Birmingham. Like many boys of his time, he found his school days
a time to be endured and his parents took him, at the age of eight, to
an unspeakable school at Boulogne where the great thrill was to
inspect the marks of recent and frequent birchings during bathing
parties. Fortunately he was removed from this establishment in 1832
by which time the death of his paternal grandfather had made the
family even wealthier than before. He was sent to a small private
school for the next four years and in 1835 to King Edward's School
in Birmingham, where he remained for three years, acquiring un-
desired learning with impositions and canings on a scale that would
gratify the most ardent floggers at modern political conferences.
From 1838–9 he worked as a medical student at the Birmingham
General Hospital and then went to King's College, London for a year.
From the college lecture rooms, he could see the lighters on the
Thames which made him yearn for travel: already, at sixteen years
of age, he had been on a tour of Europe with two young men of
twenty and twenty-two through Belgium, Germany, Switzerland and
Austria, and he had travelled with the family on various holidays.
But his journey in 1840, when he was eighteen, was crucial: on it, he
comments:[1] 'This little expedition proved an important factor in
moulding my after-life. It vastly widened my views of humanity and
civilization, and it confirmed aspirations for travel which were afterwards indulged.' In a journey lasting for several weeks, he travelled
down the Danube to Vienna, and thence to Constantinople and
Smyrna. On the return journey he had to submit to periods at two

[1] Galton, F., *Memories of My Life*, London 1908, 57.

isolation stations: at Trieste, the second of these, he was allowed to proceed by leaving behind all his clothes and swimming across a channel 20 feet wide to a quay where he acquired new, but inferior and unfashionable, clothes in which he returned home to welcoming and amused parents.

Galton's experience as a medical student spared him nothing of human misery and disease, and he overworked continuously through the years in Birmingham and London, and continued to overwork when he read mathematics at Trinity College, Cambridge. A serious breakdown in his third year prevented him from taking the Tripos and for the rest of his life he had intermittent mental and physical illnesses. When in health, however, he enjoyed himself fully and actively, and he was spared the need for constant paid work by the death of his father in 1844. Now well-to-do, he went to Egypt in 1845 and travelled widely in what is now Syria and Israel; his travels of more than a year included a journey along the Jordan to Jericho. These travels were hardly scientific exploration, but rather the activities of a sporting country gentleman, and merely part of a five-year holiday, spent partly in hunting and shooting, which lasted till 1849. Galton had acquired an interest in Africa: in 1849, he said sixty years later, 'blank spaces in the map of the world were . . . both large and numerous, and the positions of many towns, rivers and notable districts were untrustworthy. The whole interior of South Africa and much of North Africa were quite unknown to civilized man.'[1] He goes on to say that Australia was in equal need of exploration. His adulatory biographer, Karl Pearson, calls this period 'the fallow years'[2] but argues that they were of profit to Galton, as

The accumulation of experiences—however apparently aimless—is always capital of a final interest-bearing value to the man who has by heredity a receptive mind and an unusual power of storing observation. The knowledge gained haphazard in the Soudan and Syria, the pursuit of grouse on the Scottish and Yorkshire moors, the shooting of seals in the Hebrides, the observation of bird and beast, the ready presence of mind, which the hunting field encourages, the knowledge of human nature and human weakness in the gambling, wine-loving, tale-capping set of the Hunt Club at Leamington . . . were not without profit in later life. Even their value in African exploration was not to be despised. . . .

[1] Galton, F., *Memories of My Life*, London 1908, 121.
[2] Pearson, K., *Life of Francis Galton*, Cambridge 1914, I, 211.

There is little sign of serious work during this period of Galton's life, though he experimented with telegraphy on which his first publication[1] appeared in 1850: the experiments done in this investigation show the practical bent that was apparent in his travels and his later work.

African exploration

Why Galton went to Africa is not known, though the likeliest explanation is that he just wanted to do so. And it was in the current fashion. There is no evidence of exploration as a cure for the lovelorn, nor was he anxious for the fame that such an enterprise might provide. In his memoirs, he notes that his 'inclinations were to travel in South Africa which had a potent attraction for those who wished to combine the joy of exploration with that of encountering big game'.[2] Books of travel by sportsmen of the time spoke of vast heads of game on limitless grassy plains; and new routes had been opened by the travels of Livingstone and others. Through various friends and also his cousin, Captain Douglas Galton, he acquired advice from the Royal Geographical Society, which from its foundation in 1830 had been a meeting-ground for travellers, though the Society had not then developed its work of advising explorers as thoroughly as it did later, partly through Galton's interest and advice. On 5 April 1850 he set off to the Cape, where he arrived eighty days later, now well versed in the use of the sextant: he had an introduction to the Governor of the Cape from Lord Grey, the Colonial Secretary. His original intention was to go to Lake Ngami, but this proved impossible owing to the treks of the Boers. Had he gone to Lake Ngami, he might have been able to say 'Dr Livingstone, I presume' long before H. M. Stanley did, but in August 1850 he set off to Walfish Bay and began his journey across Damaraland, beyond the last missionary outposts, in country then unknown to Europeans. He saw more big game than he wanted, for on occasion lions ate his horses and mules. But he lost no men through sickness or violence: he had to break in all his stock, settle local feuds, and use 'an indolent and cruel set of natives speaking an unknown tongue'. He found women a help, though he was shocked by their propensity to change husbands at frequent

[1] Galton, F., *The Telotype: a Printing Electric Telegraph*, Cambridge 1850.
[2] Galton, F., *Memories*, op. cit., 122. The expedition is fully described in *The Narrative of an Explorer in Tropical South Africa*, London 1853.

intervals: he also awarded summary justice with a whip of rhinoceros hide. His travels led him through the territories of the Namaquas, the Damaras and the Ovampo. The Namaquas were yellow Hottentots, with a click language and a strain of Dutch blood, presumably from traders: their territories had been tenuously penetrated by missionaries. They constantly raided the Damaras, who were various Bantu tribes: these people were also addicted to fighting among themselves and few died a natural death. To the north there were the Ovampo, Negroes of a higher type, who had some fields of maize and traded with the half-caste Portuguese to the north. Their king was one Nangoro, and Galton gave him a big theatrical crown, brought from Drury Lane. Galton regarded the Ovampo as 'a charming set of niggers' but he thought the others 'brutal and barbarous to an incredible degree'. In the area of exploration, there were also Bushmen, who were nomadic hunters of Hottentot stock, and the Ghou Damup, a people living largely on roots: these people were hunted and ill-treated by the Damaras. Having completed his journey, Galton arrived back in Walfish Bay on 5 December 1851 and in England on 5 April 1852, exactly two years after his departure.

Immediate publication of the expedition's progress and work was given by the Royal Geographical Society, whose journals at that time—and for many, some would say too many, years afterwards—consisted almost entirely of the records of exploration. Indeed, part of the paper describing Galton's journey was read before his return, on 23 February, and the second part followed on 26 April. Honours came quickly, for in 1853 he was given a gold medal of the Royal Geographical Society, in 1854 a silver medal from the Paris Geographical Society and in 1855 he was elected to the Athenaeum Club under the rule that not more than nine persons could be elected annually on the ground of distinction in science, literature, art or public service. The citation of the award made by the Royal Geographical Society noted that Galton had fitted out the expedition at his own cost, conducted it successfully through the countries of the Namaquas, the Damaras and the Ovampo, and made astronomical observations of latitude and longitude. In presenting the medal, Sir Roderick Murchison made comments quotable as a period piece:

You . . . quitted a happy home, and, in the ardour of research, explored at your own cost and under great privations a region probably never before trod by civilized beings. So long as Britain produces travellers of such spirit, resolution, conduct and accomplishments as you possess, we may be assured that she will lead the way in advancing the bounds of geographical knowledge.[1]

In short, Galton was made, and he served the Royal Geographical Society for many years after his main interest had gone into other fields.

In 1853 the work of the expedition was published in book form as *Tropical South Africa*. In the same year, Galton married Louisa Butler, whom he met during a period of convalescence spent at Dover and to whom he became engaged at the Crystal Palace. Miss Butler was the daughter of the Dean of Peterborough, and came from a family of enterprise and ability, all carefully analysed by Galton in his later studies of heredity: in his memoirs[2] he notes that it is far more important to be

married into a family that is good in character, in health, and in ability, than into one that is either very wealthy or very noble, but lacks these primary qualifications. The enlargement afforded to the previous family interests through marriage is so great that much must be lost when first cousins marry one another.

After the wedding, the Galtons left for a honeymoon tour of several months, which was almost immediately followed by lengthy continental tours so that Galton did comparatively little serious work for almost two years. All through his life Galton spent long periods in European countries, but these were entirely holidays or cures at watering places for himself or his wife.

Galton was aware of the need for expeditions to be adequately provided with instruments, and wrote to the Royal Geographical Society in the summer of 1852 on this subject. In 1855 the disasters in the Crimea led him to offer to give a course of lectures at Aldershot on the art of travelling and campaigning, at weekly intervals from January to March 1856: the subjects were drinking water, fire and bivouacs, food, 'the march' including rivers and bad roads, crafts and mechanics, bush manufactures, animals of draught and burden, tents and huttings. On each of the mornings following the lectures, Galton was present to answer questions, and he was prepared to repeat the lecture of the previous evening if an adequate audience appeared. An inaugural lecture which preceded the course was published as 'Arts of Campaigning'[3] in 1855, from which it appears that

[1] *Jl R. Geogr. Soc.*, 23, 1853, lviii–lix. [2] *Memories*, op. cit., 158.
[3] Pearson, op. cit., II, 17.

he used two huts as a base for his work, of which one was a museum illustrated by sketches and models with a small library on the arts of campaigning, and the other was a workshop. Galton had a natural bent, seen from his schooldays, for improvising new tools and instruments, several of which survive in the museum of the Royal Geographical Society and the Galton laboratory in the University of London.

Ingenuity with practical common sense is characteristic of his book, *The Art of Travel*,[1] and it is impossible to resist the impression that Galton was conscious of its value as the preface is in effect a blurb. The book, he announces, includes 'the experiences, not alone of one kind of country, but those of the Bush, the Desert, the Prairie, the Water-side, and the Jungle; and the whole is arranged in a systematic manner, as a book of ready reference'. He drew heavily on the experiences of other travellers and the book is really on backwoodmanship: references to it were given in the Aldershot course mentioned above. He had the spirit of the perfect boy scout, and parts of the *Art of Travel* are much like Baden-Powell's *Scouting for Boys*. An army officer, an emigrant or a missionary should know how to use an axe, saw and chisel, a sail needle, a cobbler's awl, a blacksmith's hammer, a tinsmith's soldering iron. The twenty-five chapters are on water, fire, bivouac methods, clothes (flannel next the skin much favoured), food, discipline, defence, food storage in caches for a return journey, boats, rafts, fords, bridges, roads, the movement of heavy bodies, carpentry and metalwork, writing materials, cattle, saddlery, wagons and horses, guns and rifles, trapping instruments, fishing, medicine (gunpowder in water as an emetic). No aspect of exploration is neglected, and advice is even given on suitable presents for natives: *not* beads, he advises, as in some places the women had received so many that as they moved there was a noise like the grunting of pigs from the jangling of beads offered by previous travellers.

The work abounds in practical tips. If you wish to swim a horse across a river, drive him in, jump in yourself, seize the horse by the tail and let him tow you across. Direct the horse by splashing water in his face with your left or right hand and he—and you—will arrive on the opposite bank at the desired spot. But do not forget to hang on

---

[1] First published in London, 1855, second to fifth editions published in 1856, 1860, 1867, 1872, of which the fifth was reprinted in 1893 for the last time.

firmly to the tail throughout. Or again, should you encounter an enraged animal, keep cool, and dodge to one side or another of a bush. Should you suffer from blisters change each stocking to the other foot while on the march; and if a blister is formed, pour a little brandy or other spirit into the palm of the hand, drop tallow from a lighted candle on it, rub the feet with the mixture on going to rest—and in the morning the blister will have gone. Another hint, best demonstrated, is that you can best tuck up your shirt sleeves by turning the cuffs outside in, instead of inside out. This keeps one neat throughout the day—but why not wear short sleeves instead?

Only by reading some of the voluminous literature of the mid-nineteenth century can one capture the contemporary thrill of exploration by people who were regarded, even lauded, as trailbearers of civilization. Though Galton's *Art of Travel* may seem comical now, it was useful to other travellers: shortly after his return, Galton contributed to the famous *Hints to Travellers* published by the Royal Geographical Society. Of this long-lived work the first edition came out in 1854[1] as the work of a two-man committee consisting of Admiral FitzRoy and Lieut Raper, R.N., who wrote most of the work. It was reprinted as a pamphlet on surveying and the collection of general geographical material: in 1871, the third, revised edition was produced by a committee consisting of an admiral, a vice-admiral and Galton.[2] This was a more comprehensive work than its forerunners; though still concerned mainly with survey and the determination of latitude and longitude, it had a good deal of information on meteorological observations, instructions on the drawing of Mercator projections, several pages on the collection of fauna by H. E. Bates, then Assistant Secretary of the Royal Geographical Society and even a short note on making 'paper squeezes' of inscriptions by Rev. F. W. Holland. Galton edited the fourth edition of the *Hints* in 1880 and made it suitable for the pocket, but later editions were done by others.

The contribution to climatology

Galton's interest in climatology developed from his exploration, and from his work at the Royal Geographical Society, where he served on the Council from 1854–93 and acted as one of the honorary

[1] *Jl R. Geogr. Soc.*, 24, 1854, 328–58.
[2] *Proc. R. Geogr. Soc.*, 16, 1872, 1–78.

secretaries from 1857–63.[1] At this time he was partly responsible for editing the proceedings, and assisted with the planning of expeditions, such as that of Burton and Speke in 1856, which led to the discovery of lakes Tanganyika and Victoria Nyanza: he faced the difficulty of presenting the material in some comprehensible and attractive form. The solution obviously lay in mapping, and in 1863 he published his *Meteorographica*, or 'methods of mapping the weather, illustrated by upwards of 600 printed and lithographed diagrams referring to the weather of a large part of Europe, during the month of December 1861'. In the introduction, he explained that lists of observations in lines and columns were in too crude a state for use in weather investigations but that they became comprehensible when sorted into charts: when mapped, the essential meaning became clear at a glance. In western Europe there were over 300 observers with excellent instruments who transmitted observations three times daily to meteorological societies or government institutions, and lighthouse keepers who sent returns of wind and cloud to one of the three lighthouse boards within the British Isles: there were also many independent observers, whose co-operation Galton solicited in a circular printed in English, French and German. But no central depository of daily observation existed, and Galton notes that some countries, including Denmark, Norway and Switzerland, had no meteorological institute, and that no information was available for Sweden. Little came from France or Italy, and the bulk of the continental data came from Belgium, Holland (where Buys Ballot of the famous law was most helpful), Austria and Berlin. The data, therefore, was less adequate than Galton hoped, but wisely he went forward with the atlas.

The atlas includes a printed series of the morning, afternoon and evening weather charts at eighty stations for each day of December 1861, showing the barometric pressure, direction and force of wind, proportion of clouds, temperature and wet bulb temperature. Galton had already experimented with circles and hexagons for stations on weather maps, using a hexagon for a falling and a circular symbol for a stationary barometric pressure: if rising, the hexagon was inserted in the circle. Arrows were showing for wind on four scales, breeze,

[1] There are comments on the work of Galton at the Royal Geographical Society in Mill, H. R., *The Record of the Royal Geographical Society, 1830–1930*, London 1930.

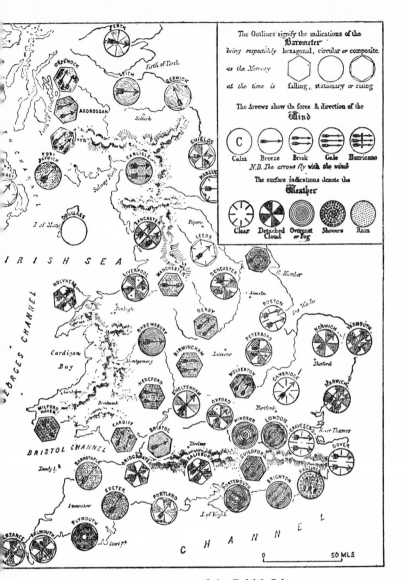

FIG. 1. Weather map of the British Isles

Labelled 'Tuesday September 3' (apparently 1861) by Galton. The various symbols originally appeared in a panel at the foot of the map. Reproduced from Pearson, K., *The Life, Letters and Labours of Francis Galton*, vol. 2, Cambridge 1924, plate VI, opposite p. 37.

brisk, gale, hurricane, with a capital C for a calm. Five grades of weather were indicated, clear sky, detached cloud, overcast or fog, showers and rain. All of this information could be included within the hexagon or circle to give a 'synchronous weather chart'. Of the maps in the atlas, the most interesting is the series showing the weather for the morning, afternoon and evening of each day in three oblongs for each time, of which the upper shows pressure, the middle one wind and rain, and the lower one temperature. One can therefore see the relationship between the various elements of weather vertically and their continuous development horizontally.

Although the theory of cyclones, or depressions, was already known, Galton conceived the idea of an anticyclone: to some extent this was demonstrated by the use of red symbols for pressures of more than 29·96 inches. Fortunately the month of December 1861 had varied weather in Europe, with a series of depressions during the early part of the month, and consequently a liberal use of black symbols: later on anticyclonic conditions prevailed with red symbols. 'As the winds in a cyclone (depression) moved in an anti-clockwise direction towards the low pressure area, so conversely in an anticyclone they moved in a clockwise direction . . .' when the wind disperses itself *from* a central area of dense descending currents, or of heaped up atmosphere, it whirls round in the same direction as the hands of a watch.[1] To Galton therefore we owe the initial recognition of an anticyclone: he also saw that the area with which he dealt was too small for effective meteorological investigation, especially when observations were so few. He was prophetic in his view that ordinary changes of wind and sky have their origin in sources far more distant than was commonly supposed, but far more data was necessary from ships, from Iceland and Greenland and other Arctic areas, as well as from tropical areas, before his 'hunch' about remote sources of weather could be confirmed as the movement and interplay of tropical and polar air. In *Meteorographica*, he included a description of the weather on the morning of Christmas Day 1861, which shows that there was the warm sector of a depression over southwest Ireland and Cornwall, and the cold sector over the rest of Ireland and England with part of continental Europe, merging into an anticyclonic area farther east in Europe. But no records were available from the Atlantic to make the picture complete. And though the initial need

[1] *Meteorographica*, 7.

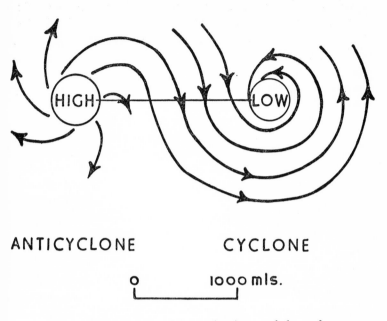

ANTICYCLONE   CYCLONE

0   1000 mls.

FIG. 2. Galton's idea of the anticyclone and the cyclone

The movement of air currents is discussed in the text: this illustration
has been redrawn from Pearson, K., op. cit., for fig. 1, 40. As noted
in the text, the recognition of the true nature of the depressions came
much later. On this see Bergeron, T., Devik, O., and Godske, C. L.,
'Vilhelm Bjerknes, March 14, 1862–April 19, 1951' in *Geophysica
Norvegica*, Oslo and Boston (Mass.), 1–25: a list of Vilhelm Bjerknes's
publications is given, 26–37. The volume, published in 1963, includes
a series of essays in commemoration of V. Bjerknes. On p. 17 it is
noted that the number of weather stations in southern Norway was
increased by 9 to 90, with the help of naval patrol boats. Bergeron
notes that Bjerknes commented (p. 18) 'During 50 years meteorolo-
gists all over the world had looked at weather maps without discover-
ing their most important features. I only gave the right kind of maps
to the right young men, and they soon discovered the wrinkles in the
face of Weather', that is the fronts. Bergeron continues by explaining
that Bjerknes' defence of the new theory of depressions, particularly
against some powerful central European antagonists, was of great
value to the Bergen workers.

was to have as many records as possible from as large an area as possible, it will be recalled that the theories of the Warm and Cold fronts in depressions were only successfully established by the detailed studies of Bjerknes, father and son, and their collaborators in

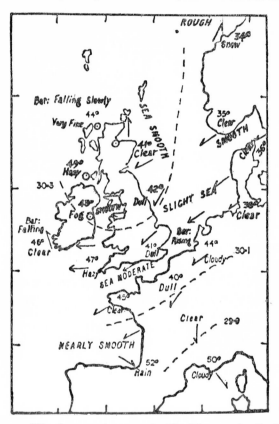

Fig. 3. The first weather map in *The Times*, 1 April 1875

The map was prepared by Galton and a comment on it was published in *Nature*, 15 April 1875. Reproduced by Pearson, op. cit., vol. 2, p. 43.

Norway during and after the 1914–18 war, when their overseas supplies of data were cut off and they wisely re-assessed the data they had already. Galton retained a life-long interest in meteorology, and was a member of the Meteorological Committee (late Council) appointed by the Government from 1868 to 1901. One advance in popular education was the publication of a daily weather map in *The Times* from 1 April 1875: in a description of the method of its preparation it is noted that

The initiative in this new method of weather illustration is due to Mr Francis Galton. . . . It is hardly necessary to allude to the nature of such charts as these as a means of leading the public to gain some idea of the laws which govern some of our weather changes.[1]

Although Galton retained an interest in meteorology, it was for others to follow the lead given in the pioneer *Meteorographica* atlas, as his main interest was diverted elsewhere. He was also interested in magnetic phenomena, and served as chairman of the Kew observatory from 1889–1901.

Other works of geographical interest

In 1855, Galton published a thirty-page paper, 'Notes on modern geography'.[2] This essay makes considerable claims for geography as a study: for example it 'is a peculiarly liberalising pursuit. . . . It links the scattered sciences together, and gives to each of them a meaning and a significance of which they are barren when they stand alone'.[3] That a synthesis of knowledge was at least in part achieved is shown in his statement that

Up to the present generation, it was not possible even for master minds to unite the scattered acquisition, to show their mutual relations one to another, and to trace the harmonious way in which all the features of the earth are organized, and how every object has its appointed post in the one mighty scheme. But now, since the writings of Maltebrun, of Ritter and of Humboldt, the case is very different.[4]

Galton thought that there was every opportunity for the map maker to use art, for example in showing various types of vegetation, and he always wanted travellers to give as vivid and vital an impression of the lands they visited, even of the smells and scents such as

that of the seaweed, the fish, and the tar of a village on the coast, the peat-smoke smell of the Highlands, or the gross, coarse and fetid atmosphere of an English town . . . the ear . . . gives . . . an individuality to every different land: the incessant and dinning notes of grasshoppers: the harsh grating cry of tropical birds, the hum, and accent of a foreign tongue, the plaintive chants with which labouring men pursue their avocations.[5]

[1] *Nature*, 15 April 1875.
[2] In *Cambridge Essays contributed by members of the University*, 79–109, published by John W. Parker and Son, West Strand, London 1855.
[3] Ibid., 81.   [4] Ibid., 82.   [5] Ibid., 98.

Galton wished to encourage geographical research by travel, and to make it easier by suggesting methods and providing instruments for travellers, to make geography a school and academic subject, and to revolutionize and humanize maps. In all these aspirations he had some success. His *Art of Travel* sold well and must have been helpful in his work for the Royal Geographical Society's *Hints to Travellers*. His main educational work was the medals scheme for schools, awarded from 1869–84 for answering examination papers,[1] and he was one of the small committee of the Royal Geographical Society which persuaded the Universities of Oxford and Cambridge to provide courses in geography. His third aim, the humanizing of maps, was done partly by precept in his own works, such as the *Meteorographica* which, though by no means cartographically perfect, gave effective visual emphasis to his scientific findings. But Galton had far more in mind and, with his permanent interest in mechanical progress, looked forward to the growth of colour lithography: he would have rejoiced in the artistic if never complete evocation of a landscape given by some modern maps. A paper of 1865[2] advocated photographs of models, to be viewed through stereoscopic lenses.

By the later 1860s Galton's original work was turning to the study of heredity, on which his first paper appeared in 1865 to be followed in 1869 by the book *Hereditary Genius*, in turn followed by *English Men of Science*, 1874, which was largely an analysis of the antecedents of certain men of science, including several Fellows of the Royal Society (not all!) who had filled in a questionnaire. Some strange geographical conclusions emerge. Far from reflecting the birthplace and origin, 'mechanicians are usually hardy lads born in the country, biologists are frequently pure townsfolk'. And 92 per cent of the scientists were born in one-half of the country, of which a map (fig. 4) is given.[3] He notes that

. . . one thin arm abuts on the sea between Hastings and Folkestone, and runs northwards over London and Birmingham, where it is joined by another thin arm proceeding from Cornwall and Devonshire, crossing the Bristol Channel to Swansea, and thence to Worcester.

[1] This scheme was only partially successful, as only a few schools entered boys for the examination. The papers were unhelpful to the advance of geography.

[2] 'Stereoscopic maps, taken from models of mountainous countries', *Jl R. Geogr. Soc.*, 35, 1865, 99–106.

[3] *English Men of Science*, 1874, 18–21.

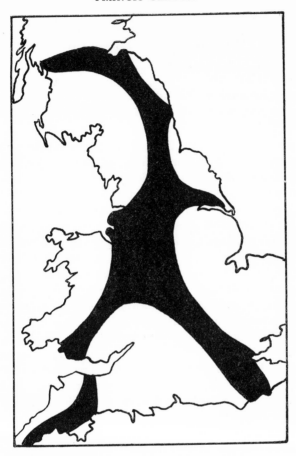

FIG. 4. Where to be born to be famous

As explained in the text, 92 per cent of scientists who became famous were born within the area shown on this map, redrawn but originally published in *English Men of Science*, 1874, 20. The coastlines are traced directly from the original map.

Ellsworth Huntington thought that there was a right time of the year to be born, see pp. 120–1.

The two arms are now combined into one of double breadth; it covers Nottingham, Shrewsbury, Liverpool, and Manchester. Above these latitudes it again narrows, and after sending a small branch to Hull, proceeds northwards to Newcastle, Edinburgh and Glasgow. Thus there are large areas in England and Wales outside this irregular plot which are very deficient in aboriginal science. One comprises the whole of the Eastern Counties, another includes the huge triangle at whose angles Hastings, Worcester and . . . Exmouth are situated.

Galton also says that 'the Border men and lowland Scotch come out exceedingly well; the Anglo-Irish and Anglo-Welsh, notwithstanding eminent individual exceptions, would as a whole rank last'. He continues with a study of the parents, which perhaps carries more conviction, and states his view that many scientists are of vigorous and even athletic tendencies. The research worker and university don addicted to climbing mountains and playing games was already a known type.

It is beyond the range of this work to explore the work of Galton in eugenics, of which he is generally regarded as the founder, and all the varied cognate studies he pursued: in passing one may note that he was in close touch with his cousin, Charles Darwin, but viewed the right course of evolution as the upgrading of the human stock by sensible selection rather than as the survival of the fittest. Like many scientists of his own later generations, Galton regarded the publication of *The Origin of Species* in 1859 as epoch-making: in his memoirs, he said that 'its effect was to demolish a multitude of dogmatic barriers by a single stroke, and to arouse a spirit of rebellion against all ancient authorities whose positive and unauthenticated statements were contradicted by modern science'.[1] He had known his cousin Charles Darwin in youth, but apparently lost touch with him for some years later, though a friendship developed from 1853 when Darwin wrote a letter of congratulation on *Tropical South Africa*: they remained good friends and occasional collaborators in experimental work to the death of Darwin in 1882. The 1853 letter includes these comments:

I cannot resist the temptation of expressing my admiration at your expedition, and the capital account you have published of it . . . I so seldom leave home owing to my weakened health. . . . I live at a village called Down near Farnborough in Kent, and employ myself

[1] *Memories*, op. cit., 287.

in zoology; but the objects of my study are very small fry, and to a man accustomed to rhinoceroses and lions, would appear infinitely insignificant.[1]

In 1862, Galton became President of Section E at the British Association as a substitute for Sir Roderick Murchison, and from 1863-7 he was the general secretary: he regarded the Association's work as of value for two reasons—first, for the discussion of new ideas before publication, and second as a sponsoring body for research committees. In 1872 he was again President of Section E, and said in his address

We are beginning to look on our heritage of the earth much as a youth might look on a long ancestral possession, long allowed to run waste, visited recently by him for the first time, whose boundaries he is now learning, and whose capabilities he was beginning to appreciate. There are tracts in Africa, Australia, and at the Poles, not yet accessible to geographers . . . but the career of the explorer, though still brilliant, is inevitably coming to an end. The geographical work of the future is to obtain a truer knowledge of the world . . . by just and clear generalisations. We want to know all that constitutes the individuality, so to speak, of every geographical district . . . to use that knowledge to show how the efforts of our human race may best conform to the geographical conditions of the stage on which we live and labour.[2]

This extract perhaps savours of the harmless but uplifting oratory characteristic of such speeches, but by 1872 Galton was seeking an understanding of psychology, heredity, and much more beyond the range of geographers. Yet he remained interested and in 1881 commented on the advance of exploration with new methods and skills, and above all improved methods of communication.

This led to the construction of 'isochronic' passage-charts, to show the extreme distances that can be traversed in 'equal times' from a common starting-point,[3] in this case London. He explained that isochrones were to be regarded as comparable with isotherms or isobars, and was apparently the first to use this term: since his day, numerous similar maps have been drawn. The assumption was that the traveller would go by the quickest means, normally by 'postal or

[1] Pearson, op. cit., I, 240. Many later letters exchanged between Darwin and Galton are given.
[2] Quoted in Pearson, *Life*, 2, 29. The address is printed in full in the *Annual Report of the British Association*, 1872.
[3] *Proc. R. Geogr. Soc.*, 3, 1881, 657-8.

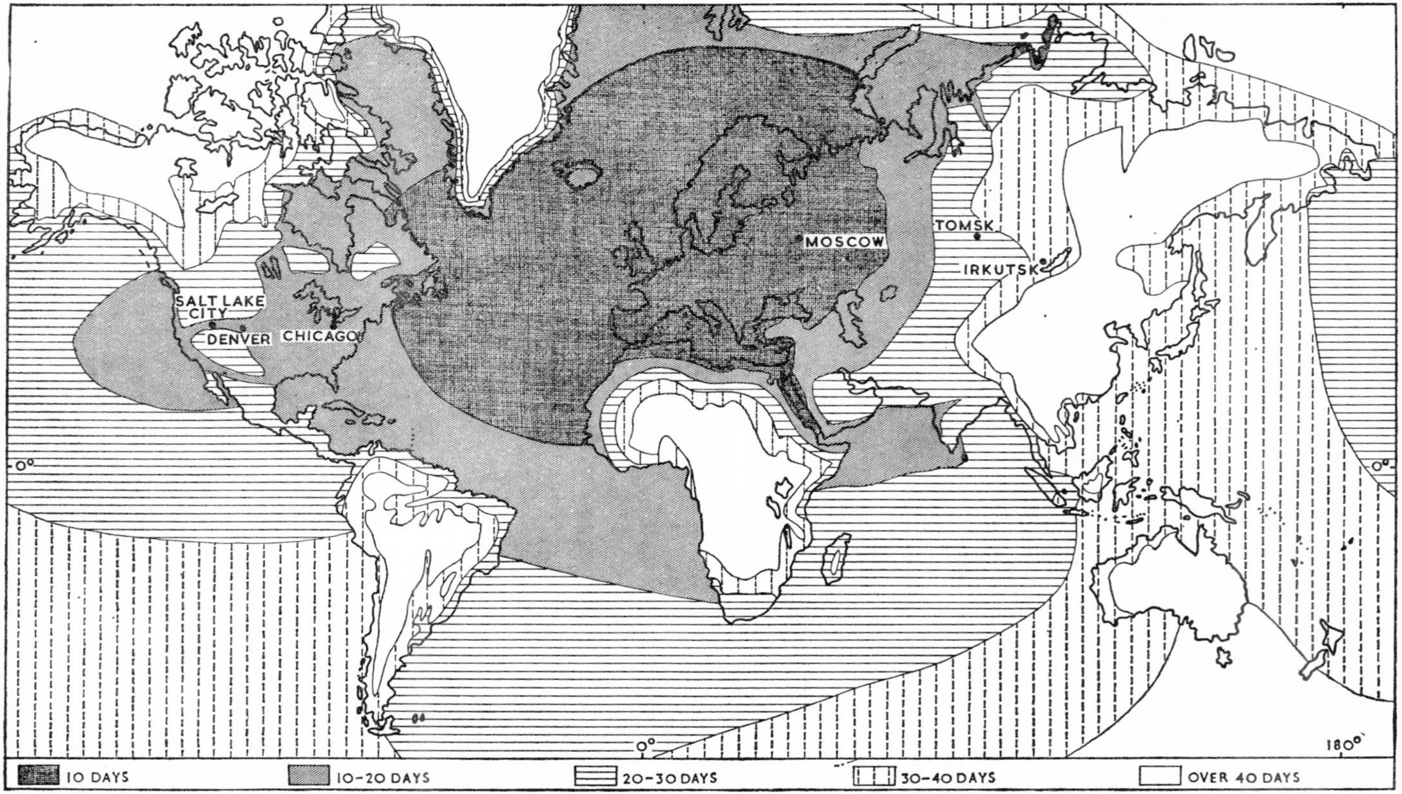

FIG. 5. An isochronous map of 1881

OVER 40 DAYS

30–40 DAYS

20–30 DAYS

10–20 DAYS

10 DAYS

other rapid regular conveyances', with such private means of conveyance as the country might afford. No delay by political upset was assumed, and it was expected that friends would be co-operative everywhere, presumably in a world having neither visas nor iron nor bamboo curtains. He used the timetables of the principal ocean lines, the estimates of the time taken to reach various places in the *Postal Guide*, private information from friends, and records of voyages. There were certain difficulties: for example a direct steamer to the mouth of the Congo would take much less time than a coastal steamer, but in this case the shading showed the quick route on the sea and the slower one on the land.

Another enterprise worthy of mention, but probably never emulated since, was the preparation of a beauty map of Great Britain.[1] He accumulated information by classifying people by looks into three classes, good, medium and bad, and carried round a piece of paper in the form of a cross to mark his observations with a specially-mounted needle. The upper end of the cross was used for 'good', the lower for 'bad' and the arms for 'medium'. Paper crosses were accumulated for each place, showing the attractive, indifferent or repellent girls seen in the streets. London was found to be the highest for beauty and Aberdeen the lowest. Apparently several studies were made at different times in some towns, and the results proved to be consistent. But Galton published no map of his findings, and perhaps it is just as well. There is earlier evidence of Galton's observation of human features in an essay of 1860, when he visited Spain to observe an eclipse.[2]

Another subject . . . with which, up to the last moment of my stay in Spain, I became no less charmed, was the graceful, supple, and decorous movement of every Spanish woman. It was as constant a pleasure to me to watch their walk, their dress, and their manner as it is a constant jar to all my notions of beauty to see the vulgar gait, ugly outlines, mean faces, bad millinery, and ill-assorted colours of the female population that one passes in an English thoroughfare. The hideous bonnet is still wholly absent in these parts and, in place of it, every Spanish woman of every class, has her dense black hair divided with a straight, clean, white parting down to the forehead, and beautifully smoothed on each side.

[1] *Memories*, op. cit., 315–16.
[2] Galton. F., 'Visit to North Spain at the Time of the Eclipse,' in Galton, F. (ed.), *Vacation Tourists and Notes of Travel in 1860*, Cambridge and London 1861.

The contribution to geography

Galton's work for societies, notably for the Royal Geographical Society, was considerable and actuated partly by gratitude for the initial impetus given to his unremunerated career by their kind reception of his work in Africa. In the whole story of nineteenth century exploration his travels, though full of perspicacious observation with some careful mapping, seem only an incident: so many went before, and so many afterwards, of whom not a few were assisted by Galton and other friendly helpers at the Royal Geographical Society. But he never went again, and never used his extensive European travels as a means of geographical investigation. Far more remarkable in many ways, if less recognized at the time, is his meteorological work, and particularly his *Meteorographica*: of interest, too, is his cartographical enterprise. But perhaps the real fascination of Galton lies in his pioneering work in psychology and in eugenics, as even a glance at a full list of his publications will show. He developed a method of investigating and classifying fingerprints; he studied the strawberry cure for gout; he devised a method of photographing several people into one composite print; he counted the brush strokes of a painter for whom he sat; he even wrote papers on 'arithmetic by smell', 'cutting a round cake on scientific principles', and 'three generations of lunatic cats'.

A great geographer of all time? No—rather a man of many interests who used his permanent leisure to good purpose, but as a geographer not of the calibre of Humboldt, Ritter, George Perkins Marsh and many more nineteenth century figures. Galton's abiding work belongs to his middle and later years, and to deeper recesses of the mind than those normally plumbed by geographers. Nevertheless he had an abiding interest in geography, though he is perhaps a supreme example of the versatile scholar of wide interests. Almost at the end of his life in 1908, he published a short paper entitled 'Suggestions for improving the literary style of scientific memoirs.' He noted that 'the memoirs published by scientific societies are blamed with justice for being more difficult of comprehension than they need be, owing to a want of simplicity in their language, of clearness of expression, and of logical arrangement'. This was due to bad grammar and syntax, the superfluous use of technical expressions that have not become naturalized among scientific men, and the lack of definition of new technical terms. However difficult the general content of a scientific paper, the introductory and concluding paragraphs should be generally intelligible to people in the same or allied sciences. He objected to the lack of training in the humanities shown by scientific men, and their inability to write essays. Of any memoir, six questions should be asked. Is it clearly expressed, free from superfluous technical words, orderly in arrangement and of appropriate length; are its new technical terms appropriate, and is the general literary style good? In this age of increasing specialization, his words may be pondered still.

BIBLIOGRAPHICAL REFERENCES

The main sources used here are the four-volume *Life, Letters and Labours of Sir Francis Galton* by K. Pearson, Cambridge 1914–24, and his own *Memories of My Life*, London 1908. The work of Pearson is marked by unlimited adulation and the *Memories* are deeply revealing of Galton's personality, and include a bibliography. The *Dictionary of National Biography* entry, in the second supplement, vol. 2, 70–3, is by Sir George H. Darwin: it pre-dates the Pearson life. The book on *Tropical South Africa*, 1853, was expanded from a paper in *Jl R. Geogr. Soc.*, 22, 1852, 140–63, and a second edition was published in 1889 by the Minerva Press, together with his article on 'A visit to north Spain at the time of the eclipse of 1860', which originally appeared in *Vacation Tours*. The *Art of Travel* was first published in 1855, and the second edition, 1856, was described as 'revised and enlarged with many additional woodcuts'. There were several later editions to the eighth in 1893. The climatological atlas, of which there is a copy in the library of the Royal Geographical Society, is *Meteorographica, or methods of mapping the weather; illustrated by upwards of 600 printed and lithographed diagrams referring to the weather of a large part of Europe, during the month of December 1861*: it was published by Macmillan in 1863. The Royal Geographical Society also has a number of his papers: the 'Suggestions for improving the literary style of scientific memoirs' appeared in the *Transactions of the Royal Society of Literature*, vol. 28, part 2, 1–7, London 1908.

On the anthropological side, see Kenna, J. C., 'Sir Francis Galton's contribution to anthropology', *Journal of the Royal Anthropological Institute*, 94, 1964, 80–93. I am indebted to Mr Kenna for comments on this chapter.

CHAPTER 3

Vidal de la Blache
A REGIONAL AND HUMAN GEOGRAPHER

FOR more than thirty years, Vidal de la Blache lived in Paris but he came from the Mediterranean and spent the last years of his life there, working to the end and fortunate in observing the development of his work by his son-in-law, Emmanuel de Martonne (1873–1955), one of the greatest geographers of his time. Vidal de la Blache was born in 1845 at Pézenas in the Hérault *département* and died in 1918 at Tamaris-sur-Mer in Var.[1] All through his working life he loved the Mediterranean and its peoples: he saw this sea as immeasurably rich in historical significance, fascinatingly varied in its littoral environments, open to influences coming from the deserts by which it is partly surrounded, assailed climatically by weather from harsher northern areas, open to the oceans of the world yet insulated from them, permanently attractive to human settlement in places yet having many areas in which any progress would be made only with the greatest difficulty, or even not at all. His vision of France was of a country having the rich background of Mediterranean civilization, flowing forwards through easy and inviting routes to cover an area that was peninsular to Europe, complex yet complementary in its agricultural resources, and not so overburdened with industry that rural life was overshadowed by the pulsating commercial life of vast towns. France,[2] he said, covers only one-eighteenth of Europe, yet within it there are areas such as Flanders or Normandy, contrasting with Béarn, Roussillon or Provence; there were areas whose affinities were with England or lowland Germany, and even with Asturias or Greece. No country of similar size, he added, had such diversities within it. Diversity of scene and of resources makes a country interesting if only because its varied products form a basis for internal trade, and it may be that Vidal de la Blache under-

[1] See the note on obituaries and general assessments of his work, end of chapter.
[2] *Tableau de la géographie de la France*, 1903, 49, published as the first volume of Lavisse's *Histoire de France*.

PLATE I. Vidal de la Blache

estimated the diversity of Britain, accentuated rather than diminished by its industrialization from the second half of the eighteenth century. But that hardly matters: though he wrote sympathetically of Britain in his *États et nations*[1] he had the Frenchman's love of France and saw it as a country of vast and enduring civilization, not possessing any marked physical barrier of separation from its neighbours and therefore different from England which, having escaped many of the devastating crises of continental Europe, intervened in wars to guard its own interests rather than through inevitable circumstances. Vidal de la Blache also referred to the maritime grandeur of Britain and (1889) noted its recent opposition to the Channel Tunnel as an example of its repugnance to any breach of its maritime frontier.

Vidal de la Blache studied history and geography, then regarded as minor subjects unworthy of serious students, at the École Normale Supérieure, where he achieved high honours in 1866. At that time geography was largely the learning of names of places and historical geography was primarily the recognition of past boundaries of states and provinces, no doubt an erudite occupation but one far from the practical field work that Vidal de la Blache developed later. Nevertheless, it was useful in the compilation of the Vidal-Lablache atlas first published in 1894[2] and still, in its sixth edition of 1951, a much appreciated work. In 1867, he became a member of the École Française in Athens, and for three years he travelled widely in the Mediterranean, with long periods of residence in Rome as well as Athens. He spent a considerable time in the Balkans, then Turkey in Europe, in Syria and in Palestine, and was present at the opening of the Suez Canal, in 1869. His first published work was a study of Herodus Atticus,[3] a rhetorician who lived from *c.* A.D. 101–*c.* 176: Vidal de la Blache admired the eloquence and pureness of expression, facility and fecundity of ideas seen in Atticus. His work on Atticus shows a clarity of writing which remained throughout his life, combined with the use of simple and effective description. In his writing he shows the superb art that appears to be no art at all.

[1] *États et nations de l'Europe, autour de la France*, Paris 1889, 263.

[2] *Histoire et Géographie. Atlas général Vidal-Lablache*, Paris 1894: later editions 1909, 1918, 1922, 1938, 1951.

[3] *Hérode Atticus. Étude critique sur sa vie*, Paris 1872. The name is given as Paul Vidal-Lablache. Another classical study, published in the same year, is *De titulis funebribus graecis in Asia minore*.

Love of the Greek classics was shown also in a review article[1] of a work on the Odyssey, which presented the view that much of the material was acquired from accounts of the Phoenician voyages, and that the scenes within it were set in real places. The geographical veracity of Homer had long been a subject of debate for Strabo regarded the descriptions of places as authentic and Eratosthenes did not. The towns, such as Ilion, Thebes and Mycene were set on isthmuses open to traders, and cruel winds, including those from the southeast of a black and tempestuous nature, assailed navigators on their journeys. Vidal de la Blache notes that Bérard's work suggested that the Odyssey was written before the time of the Greek colonization movement which did not begin until the eighth century, and adds that this is in accord with the dating of the poem by Herodotus as *c.* 850 B.C.: on the geographical evidence, he regarded Bérard's work as a fine contribution to Homeric exegesis, though many modern scholars regard it as unsound. The 1904 paper shows Vidal de la Blache's classical knowledge and it is easy to see why he was known as the 'poète Vidal' during his years at the École Normale.

The teacher and the writer

On his return to France, Vidal de la Blache taught for a time at Angers, both in a lycée and in the École Supérieure des Lettres et des Sciences, but in 1872 he went to the Faculté des Lettres at Nancy. This was a crucial phase in his life as he developed his great interest in the geography of Alsace and Lorraine, which ultimately led to the publication of *La France de l'Est* in 1917 when the question of the possible eastern frontier of France was once more acute. But the immediate significance of his move to Nancy was that it deepened the local knowledge of his country, accumulated steadily year by year through field work, much of it on foot. Perhaps it was then that he acquired the stamina that was said to astonish his students on field excursions; but other geographers of mature years have shown similar endurance since his day. His first distinctly geographical publication was an article published in 1877 on the first census of India,[2]

[1] 'La géographie de l'Odyssée d'après l'ouvrage de Mr V. Bérard', *Annls Géogr.*, 13, 1904, 21–8. Bérard had previously written articles in *Annls Géogr.*, 3, 4 and 7, and a work published by the Bibliothèque des Écoles de Rome et d'Athènes, 1894, *De l'origine des cultes arcadiens*.

[2] 'Remarques sur la population de l'Inde anglaise, avec une carte de la densité de la population', *Bulletin de la Société de Géographie de Paris*, 6e série, 13, 1877, 5–34.

in 1871. Having summarized the main statistical material, he chooses three regions different in character to show the relation between population and *son milieu géographique*. In the Ganges valley, which had clear signs of overpopulation with the constant threat of famine, the density was approximately 500 per square mile, with only 5 per cent of the people in towns of 5,000 or more. The only emigration possible was to the tea plantations of Assam, or abroad through Calcutta. In the Duabs of Punjab, the second area considered, conditions were more favourable, as there were both winter and summer rains and irrigation was already developed with possibilities of extension. Further, there were many more signs of industrial and commercial occupations. In the Central provinces between the Deccan proper and the Vindhya mountains covering more than 92,000 square miles, with over 8 million inhabitants, less than two-fifths of the total area was cultivated. Some districts were occupied by aboriginal tribes practising shifting cultivation and others by Indian peoples who cultivated large areas for such meagre returns that some wheat crops yielded only four or five times the quantity of seed sown. Nevertheless the people were not poor. Progress was most marked where the farmers were Hindus, and especially where railways had been built, and an obvious sign of progress was the increase of cotton cultivation. Everywhere in India activity was less intense than in Europe as the peasants did only a limited amount of work: in the families there were numerous children but a high death rate. Out of the general contemplation of India, Vidal de la Blache reaches the conclusion that there was a great variety of life not explicable solely in terms of natural environmental conditions such as soil and climate but partly on an historical basis, for some areas, such as the Ganges valley, had been settled for many generations, possibly for thousands of years, while others had been occupied only recently. The former were liable to famine, indicative of overpopulation, and it would be unfortunate if the more recently settled areas reached a similarly congested state. Though Vidal de la Blache never visited India, he wrote another article[1] on its census nearly thirty years later and in his *Principles of Human Geography* wove it into his writing on the movements of people throughout historical time. Vidal de la Blache saw people in an environment as living in relation to particular conditions of

---

[1] 'Le peuple de l'Inde d'après la série des recensements', *Annls Géogr.*, 15, 1906, 353–75, 419–42.

natural origin, but also as societies of varying traditions and aptitudes which had brought them, or rather by which they had brought themselves, to their existing state. With many French geographers who followed him, he was strongly aware of migration movements as instruments of geographical change and of the dangers of the population pressures that initiated such movements or might even in time result from them.

Like many pioneers of his time, Vidal de la Blache was eager to provide suitable material for schools and much of his work during the 1880s was directly educational in nature. In 1883 he published a text in physical and economic geography based partly on the history of discoveries and in the following year a book on Marco Polo appeared.[1] Apparently based largely on his teaching, his text, on *États et nations de l'Europe autour de la France* of 1889 gives an excellent summary treatment of each country with a strong historical content: though sparsely illustrated, it is marked by vivid descriptions of landscapes obviously drawn from experience. The series of wall maps long familiar to students of geography began to appear in 1885[2] and an even more enduring achievement—the publication of the Vidal-Lablache atlas in 1894—had its roots in at least ten years of solid work. But this was not all, for during the 1880s Vidal de la Blache kept his interest in erudite enquiries, one of which was a consideration of a port shown on a portolan chart, possibly at the extreme limit of the Hansa voyages and visited for salt.[3] In a well-argued paper, he reached the conclusion that it may have been Rochelle, Brouage, Bordeaux or even Lisbon. Much more important geographically was the work on regional names, published in papers dated 1885 and 1888.[4]

Of these, the earlier paper says comparatively little, though the

[1] *La terre: géographie physique et économique: histoire sommaire des découverts*, Paris 1883; *Marco Polo, son temps et ses voyages*, Paris 1884.

[2] *Collection de cartes murales accompagnées de notices*, 39 cartes, 1885 onwards.

[3] 'La Baya: note sur un port d'autrefois', *Revue Géogr.*, 16, 1885, 343–7.

[4] 'De quelques réformes dans la terminologie géographique de la France', *Revue Géogr.*, 17, 1885, 169–73: 'Des divisions fondamentales du sol française', *Bulletin littéraire*, 10 Oct. and 10 Nov. 1888. Over twenty years later, Vidal de la Blache published 'Régions naturelles et noms de pays', *Journal des Savants*, Sept.–Oct. 1909, 1–20, a review of L. Gallois' book with the same title (1908).

point is clearly made that the study of river basins could not be geographically as successful as the study of *pays* and regions. Far more light would come from a consideration of the Paris basin as a whole than of the Seine basin. In fact, the recognition of *pays* dated back to the eighteenth century through the work of geologists and, as was shown in the 1888 paper, their recognition was the living source of geography. There was an effective and vivid contrast between the Beauce, flat with vast fields and long furrows, few trees, very few rivers and all the population in villages or towns, and the Sologne, richly watered with an abundance of ponds and woods. Similar *pays*, such as those developed on Oolitic limestones, might extend across a wide area in which river valleys were only interruptions and water partings of little significance. It was reasonable to ask why the stream density, the vegetation and the density of population were so varied from one area to another, and on this basis to recognize *pays*. But this recognition began with the physical features, and led from them to the vegetation, the agriculture and the distribution of the population: then, and only then, was the possible—and frequently found—correlation with the geology to be sought. And from this enterprise, it would be possible to recognize some major divisions of France such as the Rhone-Sâone valley, one of the great transit routes of Europe with its port, Marseilles, placed at the nearest possible place to the river mouth. A geologist had commented that in their preoccupation with river valleys and basins, geographers had failed to realize the existence of the Central Plateau at all. Firmly stated in the 1888 article, the use of *pays* combined into regions was an essentially geographical contribution to learning. Two other crucial ideas also emerge in this paper. First, the *départements* were unsuitable as they were made for administrative convenience in the revolutionary period and the older provinces were also unsuitable as with exceptions such as Champagne and Brittany they were an amalgam of diverse regions. Second, geography should be regional rather than systematic only for it was desirable to aim at correlation: it was not possible to study the Normandy coasts intelligently without the chalk lands behind them, or the various promontories and estuaries of Brittany without understanding the unequal resistance to weathering of the peninsula's rocks. The initial physical correlation was followed to the final human correlation so richly shown in the *Tableau de la géographie de la France*, for which the ideas were slowly

Boundaries **not** coincident with present
département boundaries

• Regional capitals

0   200 Mls.

FIG. 6. A possible regionalization of France

This scheme by Vidal de la Blache, originally published in *Revue de Paris*, 15 December 1910, suggests a reorganization of France by distinguishing areas tributary to major provincial towns. The 'regionalist' movement was strong in France during and after the 1914–18 war: see a note in *Geogrl Rev.*, 7, 1919, 115–16, which has the above map as one illustration and, immediately beside it, a proposed subdivision worked out by the Ministry of Commerce in 1918. In the same number, the proposals of C. B. Fawcett for England are discussed on pp. 114–15 (with map) and some for Germany on pp. 116–18. But changes that seemed possible in the immediate post-war period did not happen.

evolving year by year, no doubt aided by the travelling through France during vacation periods.

Reference has already been made[1] to Vidal de la Blache's Mediterranean background by upbringing and post-graduate education. In 1886 he published an interesting paper on the effects of climate on human life in the Mediterranean.[2] Summarizing the main climatic features, he notes that there were 110 clear days a year in Naples but 171 in Palermo and about as many at Murcia: even in the north there were only a few overcast days each year, and the prevailing dryness of the air was beneficial. He uses material drawn from the climatological work of J. Hann and, on Greece, of J. Partsch, and also of James Bennett, a London doctor who settled in Mentone each winter to alleviate his tuberculosis. The Mediterranean climate had considerable contrasts between one season and another, even between different times of a single day, especially in spring and autumn. The Mediterranean peoples were apprehensive and suspicious of the weather, and strongly aware of the dangers of malaria in July and August. Indeed, they owed their vigour to the frequent changes of temperature and probably also to the cruel selection of a high infant death rate (there are hints of Ellsworth Huntington's views of a later time here, not to mention natural selection). But the bright sunshine induced a lightness and volatility of character, with a penchant to sociability and an open air life, and on a more serious plane a love of art. Buildings were constructed to provide shade when needed, and there were even covered streets. Diet was also affected by the climate, for more vegetables and fruit with less meat was consumed than in more northern climes, though in the north of the Mediterranean a more varied diet was needed and oil was plentifully used. Everywhere heavy seasonings were needed to tempt the appetite in hot weather. The concentration of people in towns arose partly from the need for mutual protection but also from the dangers of malarial lowlands worked as fertile fields till May and then left as dangerous: some peasants in Sardinia and south Corsica spent as much as half their time going backwards and forwards to their fields. Although the Mediterranean farmer was highly skilled within the limitations of his own customs and knowledge, Vidal de la Blache thought that

[1] See p. 44.
[2] 'Des rapports entre les populations et le climat sur les bords européens de la Méditerranée', *Revue Géogr.*, 19, 1886, 401–19.

conditions could be greatly improved, especially by the extension of irrigation. In this article, as in his later work, he wrote of the vast fertility developed in such areas as the *huertas* of Spain.

The Vidal-Lablache Atlas, the Annales and the Tableau

Originally published in 1894, the *Atlas Général Vidal-Lablache* was re-issued in 1909, 1918, 1922, 1938 and 1951, and is still recognizably the same work. One aim was to provide a series of maps as references for students and their teachers, and each plate is accompanied by summary notes, virtually all of which were written by Vidal de la Blache himself though revised by others and notably by E. de Martonne in the later editions. Several geographers collaborated in the production of the 1894 atlas, including L. Gallois, Camenda d'Almeida, Louis Ravenau and Paul Dupuy; and Jules Welsch of Poitiers was responsible for the geology maps. It was, in title and in scope, an *atlas d'histoire et de géographie*, for it gives a panorama of world history as well as of world geography. There are fascinating maps showing past boundaries and it is to this atlas that one turns for such information as the extent of Greek civilization, the spread of Venetian influence in the period of the medieval city-states or the extent of the Mongol empire. For every part of the world, political maps are accompanied by physical maps, which are classified further by geological, climatic and statistical maps. Having paid tribute in the preface to Carl Ritter, he notes that the ideas which led to the publication of his *Erdkunde* began with the publication of six maps of Europe[1] in 1804–6 and he states his aim clearly: to show the *ensemble des traits* which gives each country its individuality. The geology and the climate gave bases for the relief and hydrography and the physical environment as a whole influences the general distribution of population and the position of towns. In all the atlas had 248 maps on 131 plates, with an index of 40,000 place names.

By 1909, when the second edition appeared, there had been various political changes, chiefly through the spread of colonial enterprise but also through exploration, particularly in Asia, Africa and polar regions. More was known of the dune areas of Algeria, of Madagascar and of the world distribution of crops, partly altered by such

---

[1] Ritter, C., *Sechs Karten von Europa mit erklärenden Texte*, Schnepfenthal 1804–6. This was a supplement to his text book, *Europa, ein geographisch-historisch-statistisches Gemälde . . .* , Frankfurt 1804, 1807.

developments as the colonization of Manchuria by the Chinese. But in plan the atlas was little altered from the first edition to the second, nor indeed has it changed significantly since then, though in the later editions some of the maps are presented on a more adequate scale. The 1918 edition was hardly changed from that of 1909, though the maps showing the general world progress of discovery were removed, and the new political boundaries in Algeria and in the Balkans were shown—these last to be revised again shortly afterwards. Although all the editions of the *Atlas*[1] show continuity with the original, the colouring has been improved, especially from 1909: some familiar distribution lines, such as the limit of cereals and vine growing in Europe or the limit of the vine and olive in Spain still remain. There may be dangers in taking such limits as permanently fixed. At all times the cartography has been marked by a delicacy of execution which is in marked contrast to some of the cruder efforts of certain British map firms, and to the heavier, more dramatic but at times over-colourful work of some German firms.

With the *Tableau*, discussed below, the *Atlas* forms Vidal de la Blache's greatest work. Its effect on the teaching of geography in British universities was profound. On the one hand, it opened a fascinating panorama of human history as it showed the spread of civilization from one part of the Mediterranean to another, from the interior grasslands of the old world across the loess lands of Europe to the ocean fringes in Scandinavia and the British Isles or from strongholds in continental interiors to the wartime fringes as in the great medieval Mongol expansion. Many geographers loved to paint on a large canvas and give a broad panoramic view of man on the earth. Academics are prone to follow fashions in thought and from time to time the idea of 'breadth' is favoured: nothing could be broader than some of the sweepingly generalized statements on the geographical basis of history that coloured some teaching about thirty years ago, even if some people out-Vidaled Vidal. But that was not all, for with his breadth there was also depth, particularly in the discernment of local regions such as the *pays* of the Paris basin. By the time the *Atlas* was first published, the idea of dividing France into major regional units based on the local *pays* was well advanced,

---

[1] There must be few places where all the six editions may be seen together but the author has inspected them at the Royal Geographical Society in London.

E

and the excellent general maps of France were an obvious basis for the more detailed studies that were to follow later on.

Two years before the publication of the *Atlas*, the first number of the *Annales de Géographie* appeared, with Vidal de la Blache and Marcel Dubois as editors.[1] Although France had numerous journals published by the various provincial societies, as well as that of the Paris Geographical Society, founded in 1821 and the oldest in Europe, there was a need for a journal that would work to a strict geographical discipline in which material acquired from geology, meteorology and natural history must be co-ordinated and made geographical (*acclimater*): from the data of explorations the essential geographical content must be sifted for consideration and presented in a form that was at once scientific and literary rather than titillatory in its recording of curious facts and circumstances. The journal was to include, in addition to articles, critical notes, reviews, references to sources, correspondence, news and an annual summary of the progress of geographical knowledge. That the journal has admirably filled its purpose is well known: as a note in the *Geographical Review* has reminded us, the first few numbers show the broadening vision of France at the time.[2] Indeed the very first article, by P. Foncin, gives a political, social, commercial and even moral justification for the 'plus grande France', and in volume 2 there is a classic report on Madagascar, which became a French protectorate in 1885. In volume 8 F. Ratzel wrote on Corsica and two years later L. Gallois published a paper on the Patagonian Andes, finely illustrated with panoramas and maps in colour. On this the *Geographical Review* comments that 'the insert maps in these first ten volumes are the delight and the despair of the editor of today, faced with prohibitive costs of engraving and printing'. That the *Annales* owes much to Vidal de la Blache is undoubted but in 1896 E. de Martonne published a paper on the *causses* of Quercy and as the years went on he became the natural successor of his father-in-law.

Extravagant praise has been given by many to the *Tableau de la géographie de la France*, which appeared in 1903 as the first volume of Ernest Lavisse's *Histoire de France illustrée depuis les origines*

[1] The first number of the *Annales* is particularly interesting for its statement of policy and there is a notable article by E. de Martonne, 'Le Cinquantenaire des "Annales de Géographie",' *Annls Géogr.*, 50, 1942, 1–6.
[2] *Geogrl Rev.*, 33, 1943, 330–1.

jusqu'à la Révolution. It was not the work of a young man but of a deeply experienced geographer in his late fifties. It opens with the statement that the history of any people was inseparable from the country they inhabited: the Greeks lived around Hellenic seas, the English on their island, the Americans in their vast open spaces. It was his task to see how the French had lived for many generations in their own land as faithful cultivators of the soil in a country at once varied, pleasant and accessible. This could only be done by showing that the soil was of varied geological origin, and that the form of the country was the product of a comparable physical history. The first four chapters of the *Tableau* form Part I of the work, with the title 'Personnalité géographique de la France' and the phrase was apparently derived from the comment of the historian Jules Michelet (1798–1874), who said that 'La France est une personne'. But this personality is not given only by climate and soil, rather by the use made of the land by the people who lived in it. Out of this long association France had become a living entity relatively early in history: why should this happen in a country that was neither a peninsula nor an island, having no distinctive physical geography that set it apart from others? These questions are followed through four chapters, form and structure, Mediterranean influences, continental influences and the general features of France. The axis of France historically is the route from the Mediterranean to the North Sea, but much of the effective contact is, and for many generations inevitably has been, with the continental areas of Europe for there are no naturally impregnable barriers, nor on the other hand are there vast areas available for the expansion of French influence. The frontier in the east against the German lands has always been the most difficult to define: France had no opportunity of spreading into extensive open areas such as those available to Russia, China or the United States. But through the years the various regions of France, and the *pays* within them, have become welded into one whole, partly on a basis of trading exchange between *pays* whose products differ according to soil and climate. The nature of the *pays* is distinction from its neighbours but not necessarily of homogeneity within itself, for a study of the *Tableau* shows that many have a diversity of scene due to soil differences. The Brie, for example, has marls and clays in the higher eastern part, and the poor, cold soil is mainly forested, with the population residing mainly in deep cut valleys, but farther west the Brie marls are covered with limon, which long-continued drainage has made fertile, with large farms and agricultural activity varied from fruit and crop production to dairying, cattle and sheep rearing. Variety also characterizes the Beauce, for the highly permeable limestone is partly covered by limon which gives a rich and treasured soil, cultivated in long furrows without trees or ditches. Where the limestone is at the surface, the land is a desert in comparison.

The *Tableau* was written at a time when France was primarily an agricultural country, though its author's last paragraph shows that he is aware of the vast changes in progress at the time when he wrote, yet he finishes by saying that just as a storm drives waters into fury, they subside later into calmness: what is fixed and permanent in the geographical conditions of France should be regarded as fundamental. It would be easy to criticize such a statement, especially as since 1903 French life has been altered not only by over sixty years of economic change but also by the catastrophic effects of two wars: it is hardly possible to take supreme comfort from the statement that 'one generation cometh and another goeth but the earth abides'. Vidal de la Blache discusses and analyses a France that was basically agricultural, as territorially it is still even though industry has invaded many areas through planning made possible by the use of electric power and many rural areas have lost so heavily in population that their continued existence seems dubious. Yet in other works, notably his *France de l'Est*, he shows an acute awareness of industry as a transforming agent in the landscape and it seems certain that he would have watched with interest and discrimination the changes of a time later than the one he knew. It is unfair to criticize an author for failing to appreciate the nature of changes that take place after his death.

What Vidal de la Blache taught was the regional method in its fullest form. In recent years there has been a reaction from such work to the systematic approach, supposed by many who have never studied the history of geography to be 'new'. Vidal de la Blache began with the geological basis, worked through the characteristics of the climate, soil and vegetation to the agriculture, the density of population to the activities of the people as farmers, their villages and towns, the routes between them, and finally to a regional delimitation in which each unit had its individuality though each was connected with the others to make one whole—in the case of the *Tableau*, the complex if unified personality that was France. Careful study of his work will show that all through this study his basic concern was with the landscape as a whole: he saw each area in his mind's eye and kept in mind his main task—*à ne pas morceler ce que la Nature rassemble*[1]—not to divide what in nature is undivided. The aspect of correlation between what may broadly be called physical and human remains significant, but as geographical investigations become more intense studies more specialized than those of regional geography in its admittedly classic form have been favoured. And it has frequently been argued that though the methods used in the *Tableau* were admirably adapted to certain areas of France, and notably the Paris basin, they were less successful even in the south of France and not appropriate to the world as a whole: in fact Vidal de la Blache never said that they were.

The methodological problem

That Vidal de la Blache had the experience of work on small areas will already be abundantly clear: he was also concerned with the world as a whole and based his views on the principle of terrestrial unity.[2] This was first stated in an article of 1896, in which he says that the idea that the Earth is one whole, in which the parts are co-ordinated, gives geography a methodological principle of fertility and power. Nothing within the world exists in isolation, and all phenomena are interdependent. The fir tree in the garden outside the author's study is not native to England but one of its forebears was brought here from continental Europe: it sends its roots down deeply into the glacial sands and is now beginning to wither after a hundred years of existence, partly because a copper beech, whose ancestors were introduced from a more southerly area of Europe than the fir, sends its branches out a little further every year, and partly because there is a struggle for root room below the ground: before long the fir must come down as it may be no longer able to withstand the occasional gales of the winter months. Beneath the trees there are masses of daffodils that make a golden show in the spring, living only in the surface soil and unharmed by the trees above them. An analysis

[1] This famous quotation originally appeared in 'Des caractères distinctifs de la géographie', *Annls Géogr.*, 22, 1913, 299.
[2] 'Le principe de la géographie générale', *Annls Géogr.*, 5, 1896, 129–44.

of the suburban garden could show the interdependence of all life, best known to the owners by the struggle—pursued with varying assiduity—to deter weeds from ultimate conquest. This illustration was not acquired from the works of Vidal de la Blache, but from looking out of the window on a summer evening. It is applicable on the macrocosmic as well as the microcosmic scale.

In short, nothing exists in isolation and there is always a chain of cause and effect. The idea of terrestrial unity was known to the Greeks of antiquity, and derived from the mathematical demonstration of the sphericity of the earth ascribed to Eratosthenes. Ptolemy thought the same latitudes should have the same climates, the same races, the same plants and animals: at the same distance from the equator there should be elephants, rhinoceroses or coloured people. But when the world intimately known to observers was little more than the Mediterranean Sea and its borders, world influences could not be discerned clearly as local anomalies were known to be numerous. In the age of discovery, the fifteenth and sixteenth centuries, involuntary deviations from expected sailing courses led to the view that these unexplained movements of the ocean waters could only be understood when all the oceans were known. Spanish navigators along the Florida coasts spoke of *vents de retour* and expected that similar winds would occur in the Philippines, which they found in time. As early as 1620, Bacon, in *Novum Organum* noted the analogy of form between Africa and South America, and in 1650 Bernhard Varenius,[1] a North German living in Holland, then the only country still maintaining the tradition of great voyages, wrote extensively of the division of the seas, ocean movements and islands. He suggested that there should be a general geography covering the earth as a whole and the relationship of its parts and or special or local geography. During the eighteenth century the aim of making a

[1] The comment here is based on the work of Vidal de la Blache and does not purport to be a complete survey of the work of Varenius (1628–50). See Baker, J. N. L., 'The Geography of Bernhard Varenius', in *The History of Geography: papers by J. N. L. Baker*, Oxford 1963, and first published in *Trans. Inst. Br. Geogr.*, 21, 1955, 51–60. In this paper Baker notes Varenius' view that there are two kinds of geography, general and special. He quotes: 'the former studies the Earth in general, describing its various divisions and the phenomena which affect it as a whole. The latter kind, to wit Special Geography, while keeping in mind the general laws, passes in review the positions and divisions and boundaries of individual countries, and other noteworthy facts about them.'

world map was pursued and Humboldt and Ritter began their work with much preliminary investigation completed. Ritter's idea of comparative geography was by no means new: he was interested in the individuality of every living thing and to him contrasts were synonymous with life, indeed of its very essence. Areas showing vivid contrasts, between land and sea, plain and mountain, cultivated land and desert, were areas of special significance in human history, including for example Greece and Palestine. Humboldt was concerned too with co-ordination and classification, and in the Berghaus atlas of 1837,[1] pioneer maps showing the relation of climate and vegetation appeared. Ritter and Humboldt belonged to an age when the great potentialities of the world were apparent and their proclamation of the great terrestrial laws was widely acclaimed. As science became more specialized, the apparent unity was clearly demonstrated in the interdependence of climate—incidentally so fully shown when the air masses became more thoroughly known through aeronautical observation—and in the geological and geodetic unity of the globe.

When Vidal de la Blache wrote a long critical study[2] of Friedrich Ratzel's *Politische Geographie* (vol. 1, 1897) he noted that during the previous twenty-five years physical geography had attracted far more students than political geography which, though part of human geography, in its full development became wider than the study of societies and satisfying only if it achieved fullness of treatment, both physical and human. The state exists in relation to the earth, having certain qualities derived from the physical environment: mountains might offer a refuge from enemies, rivers a means of transport, islands an appropriate site for commerce but perhaps also a refuge. By the control of water and the breeding of plants as crops, river valleys had provided sites for early civilizations, but though equatorial and boreal forests had been hostile to settlement savanas and prairies could be used both for crops and animal husbandry. Man had in fact changed the earth, for example in Australia and New Zealand where the native flora and fauna was replaced, or at least sharply modified, by the introduction of European species. Within twenty years the Americans transformed their prairies into wheat fields and the bison practically disappeared. Each national group made an imprint according to its

[1] Berghaus, H., *Physikalischer Atlas*, Gotha 1837.
[2] 'La géographie politique à propos des écrits de M. Frédéric Ratzel', *Annls Géogr.*, 7, 1898, 97–111.

wishes: America, though on the whole sparsely settled, had great metropolitan areas and in Australia almost one-third of the population lived in three main cities. Vast disparities existed between one area and another: in the west of the Old World a Christian civilization was organized from the cities while in the east there was a multitude of small communities based on the villages and families. And oriental societies showed conspicuously the rift between pastoral and agricultural life. To be satisfying, all study must be local: as states France must be considered in relation to the Ile de France, Prussia to Brandenburg, Russia to the Grand Duchy of Moscow, the United States to New York. (This is effectively the *core* idea used by some modern writers.) But at the same time, no state should be studied in isolation, for each differed in its external relations: in western Europe, France had six neighbours but England was a gigantic *Thalassocratic* holding strategic points all over the world. Obviously admiring the sweep of Ratzel's group of physical and human geography, Vidal de la Blache criticized his wish to make general laws as premature and likely to introduce a dangerous dogmatism that might prove harmful.

Reviewing the position of geography in 1899, Vidal de la Blache gave a fine summary[1] of the vast increase of sources available through exploration, mapping, scientific work in geology, climatology and botany, statistical materials—among which he notes his apparently much-loved census of India. The essential impetus to geographical work was the observed difference between one country and another, and this had been so from the days of Herodotus. Why, for example, was there a sudden dramatic change of scene when crossing the 100° W. line in the United States? The accumulated climatic data showed that there were great disparities between the west and east shores of the Atlantic, latitude for latitude, and yet the people on either side had a marked similarity of *ethos*. In desert continental interiors, some people had apparently been virtually defeated through the adversity of conditions as their settlements, for example in the Lob Nor basin or the Lena valley, had been abandoned. Some ascribed such changes to climatic pulsations, notably Huntington,[2] but in such matters Vidal de la Blache was cautious and noncommittal: even if general laws could be established, they would

[1] 'Leçon d'ouverture du cours de géographie', *Annls Géogr.*, 8, 1899, 97–109.   [2] See pp. 104–23.

need modification by the observation of local circumstances such as relief, soil and climate. And he was always historically minded, for he comments that a knowledge of the geography of the past can illumine that of the present and that one can also work from the present to the past. But perhaps the most fascinating feature of his methodological writings is that he constantly urges the need for a world view yet at the same time the local scene, somewhere on the earth, appears. He never saw the world as an abstraction, but as a whole comprising millions of separate but related parts. Wherever one happens to be, there is the immediate problem: everywhere the earth and what is on it lives . . .'ce qui domine est une impression d'ensemble, de physionomie générale à laquelle contribuent le sol, le ciel, les plantes, les œuvres humaines.'

Differences between one area and another were no more susceptible of easy explanation than similarities between one area and another. In a study of 1902, Vidal de la Blache discusses the relation between geographical and social phenomena:[1] people of plains and mountains will have different agricultural products to offer, and their varying climates will be reflected in different periods of growth and harvest. A large town will be a rich market for vegetables and market gardens may become numerous, especially with horse-drawn transport. However isolated a community may be, it was noted, there will probably be some commodity such as salt which is greatly desired and not locally available, and in time a chemical industry may develop, possibly for the initial purpose of fertilizing the land: in Europe the salt trade, in one form or another, had contributed to the prosperity of Bavaria, Lorraine and Franconia. Markets where commodities could be exchanged were of vast human importance, particularly in areas such as the Sahara or in other cases where contact with other people was normally restricted and even difficult: equally interesting were the markets on the fringes of contrasting areas, such as mountains and plains or steppes and cultivated lands. In Algeria the pastoral tribes of the Tell brought their meat and dairy produce for sale in exchange for dates from the people of oases, and at such markets various implements, tools and household goods were offered for sale, many of which had been made by the womenfolk.

[1] 'Les conditions géographiques des faits sociaux', *Annls Géogr.*, II, 1902, 13–23.

Direct relations between people and the area they occupied are perhaps most obvious among primitive societies little concerned with trade and beyond the reach of manufactured goods. The nineteenth century delight in museums showing implements of various Polynesian and other tribes emphasized this point. Vidal de la Blache commented that the explorers and colonial administrators had given many examples of human societies at various stages of development and added that the study of their foods, clothes, habitations, implements and weapons showed their habits, dispositions and temperaments. The emphasis here, it should be noted, is on the people as users of opportunities, and this point is carried across many centuries of development in the comment that modern docks, elevators and machines of all types are also ethnic expressions of people at a more advanced technological stage. Though in his writing obviously aware of the idea of progress, Vidal de la Blache also recognizes that people are conservative by nature, and that progress is based on lines already marked out. They may not use more than a fraction of their resources: for example the Chinese had made a wonderful use of their soil but had never used their subterranean mineral resources and Portugal had been merely a nation of pastoral farming and horticulture before the fifteenth century, when its maritime position was exploited for the first time. The changes that have come in many nations since 1918, when Vidal de la Blache died, would have enthralled him and no doubt stimulated much interesting comment. In effect, isolation has been broken down more thoroughly than at any previous time and the technique of mass control of human beings made devastatingly efficient, due not only to compulsive ideologies but to the industrialization of vast areas such as Russia and China by the use of resources already there and for centuries unexploited. At the earlier time of Vidal de la Blache, India and China were mainly village societies, and he comments that when the British established their control of India they would have been more successful if they had recognized this rather than by accepting the principle of land ownership by great proprietors, which was 'dear' to them. Incidentally, careful reading of Vidal de la Blache's work shows that he was deeply critical of Britain.

As the work on the relation of geography to social life proceeded, Vidal de la Blache sounded many notes of caution. Even so, all civilizations were influenced by their location, the extent of the territory held, climate and agriculture which, though they engendered local diversities of life, had considerable elements in common in analogous geographical areas. The classic example of this is the Mediterranean, but equally steppes or savannas have societies which have comparable features. In tropical areas, plantations have been organized to produce cotton or fruits, with the exploitation of native labour: in Brazil, coffee farming was organized in very large units, primarily with immigrant Italian labourers, and no small proprietor could enter the market owing to the high cost of credit. This illustration suggests that the key to the situation is modern international commerce as the commodities produced must be collected, transported and exported to consumers all over the world. But to Vidal de la Blache the various forms of social living were fascinating, and he writes of the contrast between the American spaciousness and the constricted life of Europe, or of the contrast between the almost limitless plains of America, so sparsely occupied, and the thickly occupied basin of Szechwan, having a fertility developed by irrigation and fertilization over thousands of years, though placed within sight of the limits of the economically-useful world.

In a paper of 1905, Vidal de la Blache speaks of the educational aspects of geography[1] and shows once more his basic devotion to the work of Humboldt and Ritter, with his continued appreciation of the Berghaus atlas. Their fundamental ideas of relationship had made little impact in France, though É. Reclus' *La Terre* published in 1868 was partly inspired by admiration for Ritter and Humboldt: the study of geography had survived mainly through its contact with history, but as an independent study it could provide a bridge between the natural sciences and history. These ideas were developed further in a paper[2] of 1911, which opens by saying that though changes due to human activity were apparent in Europe, not least in the contemporary growth of cities, they were even more obvious in the newer countries such as the North American prairies, the pampas of South America, the Russian steppes, or parts of Algeria. From the first beginnings of settlement, changes appear. If sheep are brought in the pastures are modified: through climatic changes trees may appear where

[1] 'La conception actuelle de l'enseignement de la géographie', *Annls Géogr.*, 14, 1905, 193–207. This paper includes some material given in earlier papers, and was obviously designed for an audience of teachers.

[2] 'Les genres de vie dans la géographie humaine', *Annls Géogr.*, 20, 1911, 193–214, 289–304.

none grew before, and conversely former pastures or irrigated lands may revert to semi-desert. At all times, man was himself a powerful geographical agent but his use of resources was by no means uniform: the wretchedness of the Fuegians was in contrast to the reasonable comfort and stability of the Lapps and Eskimos, though none of the three peoples mentioned had made any significant modification of their environment. Possibly the more fortunate state of the Lapps and Eskimos was due to a social system originally evolved in middle latitudes. But anywhere life is precarious: if an irrigation canal in Sind ceases to operate, a peaceful group of cultivators may become marauding brigands. Equatorial forests inhibited progress through the impenetrable density of the jungle and the prevalence of diseases, and progress was possible only if a dry season was available, for example in savannas where some animals could be domesticated or crops grown. Man was often a destroyer of life, but he was also creative and enterprising, and introduced crops to areas where they were unknown before with conspicuous success in such cases as maize and manioc in Africa. But his choice of activities varied from one area to another: the Chinese, though so painstaking in their farming methods, never showed any great skill with trees though in the civilization of Iran and Greece the cultivation of vines, dates, figs, apricots, peaches and olives was regarded almost with veneration. Fascinated by the phenomenon of transhumance, Vidal de la Blache draws examples of it not only from the European Alps but also from the Tian Shan. Equally he shows that though the agriculturist was generally rooted to the soil, some harvests such as the sugar beet of Brie or of Saxony or the grapes of the south of France could only be gathered with the assistance of migrant labourers, then drawn from Poland, Flanders or Spain.

Throughout his work on the agricultural life of various peoples, Vidal de la Blache shows an appreciation of physical circumstances that is broad, covering not only the basic features such as plains, alluvial deposits or mountains with verdant pastures, but also their climate, initial or derived vegetation and soils. Much more is known now of the developmental aspects of vegetation than sixty years ago through ecological work: similarly far more is known on soils though Vidal de la Blache was quick to appreciate the new work as it appeared. There is no simple relationship of crops to environment. In some backward parts of the Balkans, it was noted, several types of cereals were sown in the same field more from fear that some might fail than from the agriculturally sophisticated practice of growing a mixture for fodder. In other cases, tastes changed, for example in northwest Europe where rye was being replaced by oats. In a vastly different area, Java, Hindu settlers had introduced rice though the original Sanskrit name means 'the island of barley'. Colonization of new areas by particular crops was aided by the breeding of new varieties; for example some 200 different wheats were known in the United States and more were in process of development. But from one area to another, particular crops would fit differently into the general agricultural practice: how different, for example, were the spreading wheat fields of the New World from the hedged fields of the Old World, and how different, too, was the practice of growing maize in the United States and then—in some areas at least—sending the pigs to 'hog down the corn', from its growth in Europe, with comparable sunny and wet summers, interplanted with pumpkins, beans, tomatoes or sunflowers. Northern Europe specialized in grass farming and the growth of cereals such as oats that could complete their life cycle in five months; and in this part of the world dairying for the supply of vast urban populations was obviously profitable. Possibly, it was suggested, in time a similar regimen would be developed in Canada, western Siberia and even southern Chile. Realism marked Vidal de la Blache's comment that in much of Africa areas of cultivable soil were separated one from another by stretches of sandy or granitic rock, a point made vividly in our own day by Dudley Stamp[1] in discussing the limitations of tropical areas for 'development' and reinforced by the failure of the post-1945 groundnuts scheme in East Africa. Now, as in the past, the problem is to feed the world, but even the most industrious populations may not use all their agricultural resources: for example the Japanese did not appear to have used their uplands for cattle farming, though showing continuing patience and skill in other agricultural activity. And another Japanese practice has contributed greatly to human happiness for ever since tea had been introduced in the ninth century it had been ceremonially consumed, but not everywhere in the world is the consumption of tea marked by the gracious ritual known in Japan, though

[1] Stamp, L. D., in *Our Undeveloped World*, London 1953, and in its revised and more optimistic successor, *Our Developing World*, London 1960, makes this point on the limitations of some tropical areas with vivid candour.

early in the eighteenth century in England it inspired Alexander Pope, in 'The Rape of the Lock' to write of the Queen

> Here Thou, great *Anna*! whom three realms obey,
> Dost sometimes Counsel take—and sometimes *Tea*.

Vidal de la Blache caps his reference to tea by a reference to Thucydides' comment that superfluities are of the very essence of civilization.

Some writers on Vidal de la Blache have drawn particular attention to his view that geography is fundamentally the study of places[1] rather than of men but this statement should not obscure the fact that he sees every place complete with its inhabitants, and not only with those now present but with those who have gone before, making the place what it is through the patient toil of centuries. Apart from the vast changes wrought by man in his efforts to make and re-make the earth, he has undoubtedly been influenced by world changes beyond his control, such as the considerable oscillations of climate from the Quaternary era onwards:[2] on these, as on many other matters, much has been done since his day. Fecundity of ideas obviously marked the work of Vidal de la Blache, and for many years before his death in 1918 he had the hope of producing a general human geography of the world, eventually published posthumously under the editorship of E. de Martonne. Apparently an initial memorandum on the possible scope of the book was sent to the publisher, Max

---

[1] The quotation as given in Hartshorne, R., *The Nature of Geography*, Lancaster, Pa., 1939, 241 (417) is 'Geography is the science of places and not that of man: it is interested in the events of history insofar as these bring to work and to light, in the countries where they take place, qualities and probabilities that without them would remain latent.'

If only the first eleven words are quoted, virtual dehumanization of geography would appear to be conceded by Vidal de la Blache, but the whole article of which it is a part is suffused with the concept of the living world rather than in some inanimate concept of 'place'. Hartshorne, in *Perspective on the Nature of Geography*, Chicago and London 1959, 56–7, has noted that though Vidal de la Blache was concerned with the man–nature relationship, as a student of history he 'felt it necessary to emphasize, more so than Ratzel, the elements of uncertainty in the man–nature relationship that results from differences of cultures, *genres de vie*, as well as from decisions of individual men'.

[2] 'Des caractères distinctifs de la géographie', *Annls Géogr.*, 22, 1913, 289–99. The 'places' quote used by Hartshorne and others comes at the end of this paper.

Leclerc, as early as 1905.[1] The work is in three parts of which the first, on the world distribution of population, was in effect complete and indeed already published in the form of articles in the *Annales* for 1917 and 1918, though de Martonne was of the opinion that it was intended to add a chapter on America to those on Africa, Asia, Europe and the Mediterranean with the general survey which are published.[2] Parts II and III, however, on 'elements of civilization' and on transport were mainly in the form of notes and first drafts except for two or three chapters virtually ready for publication. It may well be that de Martonne knew more of the possible final scope of the book than he reveals in a markedly frank preface. Clearly there was to have been a full treatment of cities, for fragments of a section on this subject were discovered and printed. In recent years the advance in urban geography has been so rapid that many may forget that some of the earlier advances were made in France, such as the publication of Raoul Blanchard's work on Grenoble,[3] in 1912, of which Vidal de la Blache was undoubtedly aware. No doubt a modern urban geographer would regard such a work as naïve, excessively historical in approach and sadly deficient in mapping evidence, especially on the statistical side. Even so, it was a beginning. To a great extent, Vidal de la Blache saw towns in relation to agricultural life, and readers of his human geography will doubtless remember his fine exposition of some Mediterranean towns populated largely by those who went forth to the fields around them day by day to labour, retreating homewards at night to sites that gave mutual protection against malaria or other diseases and in an earlier time against marauding warriors.

The impact of Vidal de la Blache's writing on geographical thought was greater than many people appreciate. Deriving his ideas partly from the classical German sources, he interpreted them afresh, and

[1] Any publisher who reads this may draw the comforting reflection that some of the aspirants who bring him plans for books may produce a masterpiece many years later.

[2] *Principes de géographie humaine*, ed. E. de Martonne, Paris 1921, was translated by M. T. Bingham as *Principles of Human Geography*, Chicago and London 1926. The various articles used as chapters in *Principes* appeared in *Annls Géogr.*, 26, 1917, 81–93, 241–54, 401–22, and 27, 1918, 92–101, 174–87.

[3] Blanchard, R., *Grenoble*, Paris 1912, deals with the river, the site, the history (a large part of the book), industries and the town as a regional capital.

drew on his own varied experience to illustrate them and to develop new ideas from them. From the background of nineteenth century scientific discovery, especially in the natural sciences, he drew his conception of terrestrial unity in which man was the greatest agent of geographical change, linked by many strands to his environment yet able to change the environment by drawing on its inherent, if not necessarily readily discerned, resources. His classical background and his wide knowledge of history gave him a breadth of view that is staggering to modern workers, inevitably more specialized by training and generally also by inclination. Some British geographers of an earlier time regarded his work as supreme: P. M. Roxby (1880–1947) once told the author that in Liverpool University students in the first year read, in French, the *Principes* with extreme care, line by line and word by word, in short much as a student of literature would study a text of Shakespeare or Molière. Excessive veneration perhaps ? The essential point, which may not be clear to those who were not born by the interwar period, was that progressively minded thinkers in many fields worked forward to a time when the scientific unity of the world would be crowned by a human unity of all peoples without which, they argued, in the end all civilization must perish. Vidal de la Blache's work offered a discussion of human life in its various activities set in a wide variety of physical environments, differing widely from one place to another even if the initial possibilities seemed comparable. Probably the work appealed most to those who had a deep acquaintance with rural life as the work on cities is less thoroughly developed than the survey of agricultural activities. Vidal de la Blache was fascinated by the disparities of population density, and commented that 'the existence of a dense population—a large group of human beings living together in the smallest space consistent with certainty of a livelihood, for the entire group—means . . . a victory which can only be won under rare and unusual circumstances'. Man, himself an end-product in the process of evolution, had expressed himself in a wide variety of ways, the study of which, many thought, must lead to a greater sympathy with and understanding of the world as an inevitable whole.

The final work

What has been written above will seem utopian to the point of foolishness to many readers faced with a world in which divisions are all too apparent. It was the misfortune of Vidal de la Blache that the last years of his life were spent in a time when all the French civilization that he revered seemed permanently threatened by Germany, but his last work, partly published in the *Annales,* was a critical study of *La France de l'Est,*[1] in effect a plea for the return of France to a frontier on the Rhine. For thirty years he had meditated on this question, and this book was therefore not something hastily produced but rather the result of long and careful thought. Alsace and Lorraine were conquered by French ideas from the seventeenth century and welded to France in the revolutionary period. The spread of Prussia from its origins across the sandy lowlands of Brandenburg gave Europe a people who were hard working, anxious for national unity and willing to sacrifice much for it, but at times violent and therefore dangerous. Alsace and Lorraine under French rule could open in Europe a way for the penetration of liberal influences, for the basis of civilization was the right of any people to have their moral and economic independence. The inclusion of Alsace and Lorraine in France would open useful routes from the Rhine to the Mediterranean and the English Channel.

Nevertheless, as Vidal de la Blache had shown at various times in his earlier writings, the settlement of the eastern frontier of France was not easy. To some extent 'l'Est' was an entity apart, its physical environment sharply different from that of the Paris basin, which had not only a natural geographical unity but also a human unity, developed partly through the cultural, administrative and commercial impact of Paris. In Lorraine and Alsace local differences were marked: each valley and basin had its individuality, even its differences of dialect and the interpenetration of forests and farmlands

---

[1] *La France de l'Est* was published in Paris in 1917 and reached its third edition in 1919. The date given at the end of the preface is December 1916. Articles in *Annls Géogr.,* 25, 1916, 97–115, 161–80, 'Évolution de la population en Alsace-Lorraine et dans les départements limitrophes' are based partly on maps which also appear in *La France de l'Est.* Under the two names of Vidal de la Blache and L. Gallois, *La bassin de la Sarre, clauses du traité de Versailles, Étude historique et économique* was published without date but probably in 1919. This included an article by Gallois in *Annls Géogr.,* 38, 1919, 280–92, and also a memoir written in 1917 by Vidal de la Blache *La Frontière de la Sarre d'après les traités de 1814 et de 1815* for the Comité d'études présidé par M. Ernest Lavisse. L. Gallois laments the fact that Vidal de la Blache died without having the joy of seeing France's return to the Rhine, which he regarded as essential to European peace. The July 1919 number of *Annales* is a study of the Treaty of Versailles.

added to the variety. Even so, the prevailing culture, notably of the towns, Metz, Nancy, Strasbourg and Colmar, was essentially French, despite the German *patois* spoken in many areas, and the general purpose of treaty adjustments in the eighteenth century, as in 1815, was to give France a Rhine frontier. As the nineteenth century wore on, railways, improved canal and river services, with the industrial expansion based on coal, iron and salt, brought large new industrial communities into existence, while the agricultural population diminished. From 1871, when Alsace and Lorraine were added to Germany the whole policy was to induce trade with areas to the east rather than with France and to turn the whole area into a part of the Prussian state regardless of the wishes of the population. The book contains both a detailed historical as well as a geographical analysis of the situation. In fact Vidal de la Blache did not live to see what happened at Versailles, for he died on 5 April 1918, his death hastened by the loss in 1916 of his son Joseph, fighting for France in the war.

BIBLIOGRAPHICAL REFERENCES

Other works of Vidal de la Blache

At various times Vidal de la Blache wrote short articles in non-professional journals as well as notes and reviews in the *Annales*: the major works are listed under the references above. Two papers of the 1890s are worthy of mention: 'Les voies de commerce dans la Géographie de Ptolémée', *Comptes Rendus de l'Académie des Inscriptions et Belles-Lettres*, Paris 1896, which is illustrated by a map from the *Atlas* showing the economic state of the Greco-Roman world; 'Notes sur l'origine du commerce de la soie par voie de mer', same journal, 1897, in effect a study in historical-economic geography at c. A.D. 2. Another erudite paper was *La Rivière Vincent Pinzon. Étude sur la cartographie de la Guyane*, published by the Université de Paris, Bibliothèque de la Faculté des Lettres, 1902. This is a fascinating study of a frontier defined at the Treaty of Utrecht in 1713 between France and Portugal which had involved prolonged argument between the two powers and from the liberation of the South American colonies between France and Brazil. The work begins with the discovery of the area, the various early maps, and shows that the river had been so changed in its form by coastal erosion and the emergence of new channels that a modern compromise solution must be reached.

That Vidal de la Blache conceived the *Géographie Universelle* with L. Gallois is known to all its readers, for his name appears on every volume. In fact the first of the twenty-three volumes did not appear until 1927, when the British Isles and the Low Countries were treated by A. Demangeon, and the last volume on France, also by Demangeon, came out in 1948. Ironically enough Demangeon (1872–1940) was a victim of the second world war.

Obituaries and appreciations

The main obituary is by L. Gallois in *Annls Géogr.*, 27, 1918, 161–73: others include R. Blanchard, *Recueil des travaux, Institut Géographique Alpine*, 6, 1918, 371–3, G. G. Chisholm, *Scott. Geogr. Mag.*, 34, 1918, 226–7 and shorter notes, *La Géographie*, 32, 1918–19, 369 and *Geogr. J.*, 52, 64–5. The place of Vidal de la Blache in French geography is discussed in the *Geogr. Teach.*, 9, 1918, 195–6, 201–5. A review of his work to 1910 with a bibliography is given in *Geographen-Kalendar*, 8, 1910, viii–xxx by L. Gallois and E. de Martonne. A study compiled to celebrate the centenary of Vidal de la Blache's birth appeared in *Acta Géographique, Comptes rendus de la Société de Géographie de Paris*, no. 4, July–October, 1947, 2–6. A survey of the French School of geography by R. J. Harrison-Church appears in G. Taylor (ed.), *Geography in the Twentieth Century*, London 1950, and there is a summary treatment in G. R. Crone, 'The men behind modern geography. Paul Vidal de la Blache and geography in France', *Geographical Magazine*, 23, 1950, 129–31.

CHAPTER 4

Jovan Cvijić

A RELUCTANT POLITICAL GEOGRAPHER

RARELY in human experience can more have developed from a man's long walks through the countryside than from the summer wanderings of Jovan Cvijić. Some men have found that long walks through mountains give opportunities for thought which are hard to acquire in the daily pressure of duties: some men have gone forth alone, only to find themselves drawn into the lives of hundreds of country people as a welcome visitor giving them a breath of a wider world which they are eager to acknowledge with gracious and even charming hospitality. Cvijić had both these experiences, and much of his work on the human side of geography was based on his knowledge of the peoples of the Balkan peninsula he knew as friends. It has been a criticism of much past and not a little present human geography that it appears to deal with everything except the people: Cvijić was carefully observant of what has uneuphoniously been called 'material landscape features', and has illustrated his works with drawings and photographs of house types; but he did not stop there. It may be that he went further into a study of the ethnographic and the psychological aspects of life than his successors would venture, but since his day both ethnography and psychology have advanced rapidly as studies having their own expertise and methods. It was not primarily as a student of human nature and backgrounds that Cvijić went through the Balkans, but rather as a searcher for explanations of their physical features. To a great extent the Balkans were a geographical *terra incognita* in the years before the working life of Cvijić, but he made them well known and had the embarrassing success of planting several Serbo-Croat words firmly, it seems almost irrevocably, in the literature of geomorphology, at least for limestone landscapes.

Yet all his great achievement as a human and a physical geographer seems minor beside his most permanent memorial—a share, probably a great share, in the addition of Yugoslavia to the map of

PLATE 2. Jovan Cvijić

Europe. One underlying cause of the 1914–18 war was the impending disintegration of the Austro-Hungarian empire. Following the withdrawal of the Turks from almost all of the Balkan peninsula, the fights between various new national groups in the years before 1914 showed that stability could not be achieved there without great strain, but many pages of Cvijić's work show that in spite of all the evidence to the contrary, he still believed that some form of Slav unity was possible. Shrewdly, he recognized that the Bulgars were a distinct group, different from the Slavs and better served by the possession of some territory of their own than by inclusion in some new pan-Slav unit. No one saw more clearly than Cvijić the complexity of the Macedonian problem: at all times he was ready to see the intricacies of any situation. Like many geographers of his time, particularly the French, he was strongly aware of the effects of migrations of people from one area to another; and in the Balkan peninsula, subjected to a history of constant stress, these had been numerous. They were not, however, easy to distinguish, and many of them were elucidated by enquiry from groups of people. Political geographers are prone to think in large terms (often too large), but Cvijić built up his major structure brick by brick, day by day, observation by observation, and so made it vivid and convincing. In his later years, he turned again to his loved geomorphology, and it is perhaps as well that he did not live to see the mounting years of crisis in the 1930s and their tragic outcome in his own country, destined to rise again after 1945.

Cvijić was described by the French geographer, E. de Martonne, as *un géographe serbe, un géographe balkanique, un géographe européen, un géographe mondial.*[1] Such French eloquence aptly describes the career of a geographer who, beginning his studies round the small town of Loznica in west Serbia, where he was born on 6 October 1865, gradually extended his work from his native Serbia to the entire Balkan peninsula, and so to the place of the Balkans in Europe as the home of independent nations able and eager to work out their own destiny, and finally to researches, particularly in physical and regional geography, that would be read by the discriminating all over the world. But by the discriminating only, for

---

[1] Quoted by B. Ž. Milojević in preface to J. Cvijić, *La Géographie des terrains calcaires*, Monographie, Tome CCCXLI, Classe des sciences mathematiques et naturelles, Belgrade 1960, vii.

La péninsule balkanique never achieved a second edition. From another source comes a generous recognition of Cvijić's work. H. W. V. Temperley,[1] in a section on the claims to territory made at the Versailles peace conference of 1919, notes that

the memorandum of the claims of the Serb-Croat-Slovene state . . . was . . . like the rest, based upon considerations partly ethnic, partly historic, and partly strategic and economic. But it differed from almost all the others, as, for example, the Italian, the Greek and the Rumanian, in that the ethnic argument was the strongest and most permanent element on which the Yugoslavs could rely . . . [it] bore obvious traces of the hand of M. Cvijić, the most learned and enlightened not only of Serbian, but of all Balkan geographic experts.

As a Serb geographer, Cvijić's first and possibly most enduring work in geomorphology was on limestone landscapes and, to a less extent, on the evidences of glaciation in the Balkan peninsula. The work on limestone karsts and other phenomena has been quoted and emulated everywhere, and even the Serb terms are generally known in the world literature of geomorphology. Probably this was his greatest interest, for there is some evidence that he returned to it in his last years, by which time it had stimulated similar work in many parts of the world under a wide range of climatic conditions. The work on glaciation was, like the limestone studies, of considerable importance, but naturally enough had a less wide appeal. Cvijić's human and regional studies of the Balkans made him a European figure, which indeed he had become already through his geomorphology; and he became a world figure partly through the accident of historical events which made his work in human and regional geography a cogent argument for the addition of Yugoslavia to the map of Europe. But the laurels were won also for his geomorphology. To anyone beginning a geographical career, the study of Cvijić's life shows that long-continued and doubtless arduous field work can bring to light features of universal significance, which is perhaps a sobering reflection in days when mathematical and statistical methods are regarded by some as a final key to geographical research. Another reflection is that in his writing of human and regional geography, Cvijić cast his vision widely over people's activities and views, but in this he was in line with many writers of his time. As an advocate, Cvijić was the more effective for his restraint and

[1] *A History of the Peace Conference of Paris*, London 1961, vol. 4, 207.

reasonableness: he never pushed his case beyond the bounds of apparent logic and his prejudices, or at least his understandable human feelings such as a dislike for the Bulgars, are expressed in a temperate manner, or at least in terms more of pity than of anger. He was fortunate in possessing, at Versailles, the cast of mind that was understood at the time: his deep belief in self-determination for all peoples possessing at least a nascent concept of nationality, and his condemnation both of inefficient Turkish rule and of misguided Austro-Hungarian attempts at imperialism in the Balkans were acceptable to the delegates. But he may well have seen that the achievement of democracy and national cohesion in a large Balkan state would be a matter of some difficulty, for his works show a full awareness of the diversity among the peoples in various areas, not to mention the formidable natural barriers to the development of a modern communications system.

The pattern of Cvijić's career as a geographer was a beginning on the physical side with a gradual deepening of interest in the human aspects of study culminating in the widely-spread themes of his *Péninsule balkanique* of 1918, written when he was over fifty years of age. In his later years, after his return to Belgrade, he followed his original bent towards research in physical geography: at all times he appears to have been singularly modest about his larger enterprises, notably in the preface to *La péninsule balkanique*, where he explains that he was mainly a physical geographer but that during the Balkan wars of 1912–15 he was forced to consider questions of political and human geography. A study of his works, however, shows that the interest in people was strongly developed before 1912, partly due to his annual travels, normally lasting some weeks, in the Balkan peninsula from 1888 onwards. Stimulus to the human studies came from Vidal de la Blache, who received Cvijić at the Sorbonne during the war years and arranged for him to lecture:[1] not unusually a book may have its trial expression in a lecture course and the *Péninsule* was based on a course. Significantly, Cvijić criticizes Ratzel and Brunhes as human geographers as he thought they excluded man from 'human' geography to an unjustifiable extent by concentrating mainly on the material landscape features, to use the modern

---

[1] In the *Bulletin de la Société Neuchâteloise de Géographie*, 26, 1917, 172–3, the then president, Gustave Jéquier, notes that Cvijić had been a refugee at Neuchâtel before he went to the Sorbonne.

American idiom. A senior American geographer, G. T. Trewartha, has said much the same thing in our day.[1] No doubt, to quote a well-worn saying, 'the proper study of mankind is man' but any geographer in the 1960s who wrote with the width of Cvijić would be regarded as spreading too wide a net and trespassing on specialisms that since 1918 have achieved their own distinction of study, though at his time Cvijić was by no means unique in his views. His interest in physical geography was permanent, and as it developed earlier it will be convenient to consider it first.

The physical geographer

Cvijić acquired his first impressions of the karst country in his home district. He had the good fortune to study in Vienna with Albrecht Penck (1858–1945) and Eduard Suess (1831–1914), who helped to extend his interest from karsts to glaciology and limnology, especially the shorelines of lakes: clearly the glaciation of the mountains in the Balkan peninsula was a study worth pursuing not only for itself but also in relation to the general glaciation of Europe. Cvijić's first paper was published in 1888 and dealt with the movement of underground water in limestone areas: in time a great deal of speleological work was done, some of it through the enterprise of the Geographical Society of Belgrade. In 1893, Cvijić published a paper on karsts[2] which put forward his main theories of the development in limestone landscapes. Assuming that the limestones were 99 per cent soluble, the *terra rossa* developed from the insoluble residue of clay and iron oxides: *terra rossa* not only absorbs water easily but retains it, and provides some admirable land for agriculture. The evolution of a karst landscape proceeds largely by solution, and its development may bear no relation to sea level as the saturated zone provides a base level for erosion, which could be either above or below sea level. Effectively the base level was associated with the water table, or aquifer, which might be either a bed of impermeable rock such as an underlying clay, or else a layer of material cemented into impermeability by the deposition of calcareous material known as tufa, though possibly this might in the end waste away by the later

---

[1] Trewartha, G. T., 'A case for population geography', *Ann. Ass. Am. Geogr.*, 43, 1953, 71–97.

[2] 'Das Karstphänomen', Penck's *Geographische Abhandlungen*, Vienna, Band 5, Heft 3, Vienna 1893.

percolation of water. The main theories of Cvijić on the evolution of the karst landscape have become standard, and are widely reproduced in geomorphological textbooks. Twenty-five years after the 1893 paper was published, Cvijić published in French a paper which used his own and other people's studies:[1] still convinced that the key to the situation lay in the circulation of water, he argues that karst landscapes may develop in several ways, but that some form of vertical erosion to the underlying impermeable layer was inevitable. This 1918 paper was illustrated by some fine drawings.

Although Cvijić was mainly concerned with limestone landscapes to 1895, he kept his interest in them throughout his life and from 1894 to 1914 sent numerous communications on caves and underground channels to the Bulletin de la Société de Spéléologie, Paris, to the Serb Geographical Society and to the Academy of Sciences in Belgrade. But from 1897 Cvijić wrote papers on the glaciation of the Balkans, from 1899 on the geology and tectonics of the Balkans and from 1901 on the relief of former seas and lakes of the Tertiary period. Much of the material is in Serb in the proceedings of the Serb Geographical Society, but some was published in German. At the 1903 International Geological Congress at Vienna, for example, a paper[2] was presented which included some basic geological material and a coloured structural map showing different types of folds, fault zones, crystalline masses, horizontal sedimentary rocks and other features from the west side of the Morava valley to the Danube, and across to the Black Sea. A geological atlas[3] published (in Serb, French key) in 1903 for Macedonia and Old Serbia included maps showing the geological strata with considerable attention to the igneous and metamorphic rocks and also to superficial deposits: other maps included the fold and trend lines, reproductions of various old maps from Ortelius onwards and three pages of geological sections. Five years later, in 1908, Justes Perthes of Gotha published

---

[1] 'Hydrographie souterraine et évolution morphologique du karst', *Recueil des travaux de l'Institut de géographie alpine*, Grenoble, 6, 1918, 375–426.
[2] 'Die Tektonik der Balkanhalbinsel mit besonderer Berücksichtigung der neueren Fortschritte in der Kenntnis der Geologie von Bulgaren, Serbien und Makedonien', *Comptes Rendus IX Congrès Géologique International de Vienne 1903*.
[3] The Serb title means 'Geological Atlas of Macedonia and Old Serbia', Belgrade 1903.

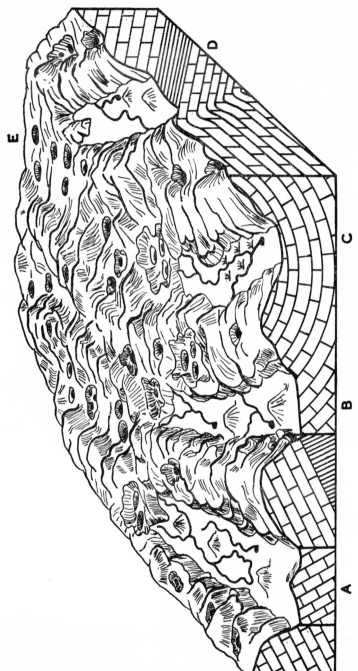

FIG. 7. Landforms in a Karst area

Cvijić's maps[1] of an area extending north and south of the Danube west of Timok, on the 1 : 200,000 scale showing erosion platforms, with various recent deposits and on the 1 : 300,000 scale showing igneous deposits and stratigraphy, with faults, dunes and moraines. A short article of 1908[2] shows that Cvijić was following with close interest the recent work of E. de Martonne on the southern Carpathians, including his suggestion that various peneplanes existed, and that the ideas of W. M. Davis on physical geography were respected widely at that time. Much of Cvijić's work on physical geography was essentially of a pioneer character, and it was a useful basis for the regional work that was eventually to follow.

When Cvijić first put forward his views on glaciation, it was generally believed that there had been no glaciers in the Balkan peninsula. The researches of geologists, notably Ami Boué in the early part of the nineteenth century, gave no hint of a glaciation and later workers, from 1870–90, all agreed that the Balkan mountains have never been glaciated. In 1890, however, as a young man of twenty-four, Cvijić climbed Char-dagh, then thought to be the highest mountain in the Balkans, and around its summit, Ljubotin, saw three cirques, one of which had a lake. But as there were neither moraines not striations visible, Cvijić decided that it had not been subject to glaciation, though in the same summer and in 1891 he found many traces of permanent snow on sunny mountains in Rila

---

[1] *Entwicklungsgeschichtliche Karte des Eisener Tores, I : 200,000*, Gotha 1908.
[2] *Petermanns Mitt.*, 54, 1908, 114–17, 'Peneplains und epeirogenetische Bewegungen der Südkarpathen'.

Fig. 7. Cvijić produced a large number of block diagrams, of which this one shows an area of structural complexity. Four stages of *polje* development are included:

A—a *polje* coinciding with a doubly faulted depression
B—a *polje* coinciding with a faulted fold
C—a *polje* coinciding with a syncline
D—a *polje* formed by erosion and solution along an outcrop of non-calcareous rock
E—a high *karst* platform, pitted with solution hollows of various types.

Originally published in Cvijić, J., 'Hydrographie souterraine et évolution morphologique du Karst', *Recueil de l'Institut de géographie alpine*, 4, 1918, 409: reproduced in N.I.D., Geographical Handbook, *Jugoslavia*, 1, 1944, 51.

and Durmitor and reflected that with only a slight fall in temperature glaciers might develop. Steadily the evidence grew: in 1896 he found numerous traces of former glaciers in the Rila of Bulgaria, and in 1897 he observed similar traces of glaciers in the Dinaric mountains of Bosnia, Hercegovina and Montenegro. On the Rila in 1896 there were several hundred snow patches, of which a number covered 700 square metres and the largest, of 1,000 square metres, was 8 to 12 metres thick and almost in the plastic state of névé. The snow line was probably at 3,050–3,080 m., some 130–160 m. above the highest summit. Although there was some mention of glaciation in a paper published by Cvijić in Vienna in 1890,[1] the first publication came ten years later in Paris[2] when he noted that in 1899 he had been on a geographical excursion through Bosnia, Hercegovina and Dalmatia with A. Penck and W. M. Davis (1850–1934), both of whom agreed that there were traces of glaciation on the Bjelasnica near to Sarajevo and on the Orjen near the Gulf of Cattaro: they suggested that in the glacial epoch the snow line was at 1,800 m. on the Bjelasnica and at 1,400 m. on the Orjen. In the 1917 papers these figures were modified and the heights suggested for latitude 42°, from west to east, were 1,300 m. on Orjen, 1,450 m. on Sinjajevina, 1,550 m. on Pnokletije, 1,740 m. on Char-dagh, and 1,850–1,880 on Rila: presumably conditions during the glacial period in the Balkans were similar to those of Norway some fifty years ago (but not recently as there has been a climatic amelioration). Also in the 1917 paper Cvijić notes that the Rila, the hydrographic centre of the peninsula, having the sources of the major rivers, has a generally rounded form except where glaciers developed on the north and east sides, with cirques on the south and west. In all there were over thirty cirques and more than a hundred lakes that were either moraine-dammed or in cirques, but most of the glaciers had at some time flowed out of the cirques into the valleys. The moraines are tentatively related to various halt stages of the glaciation in the Alps: beyond them glacifluvial deposits occur. But the finest evidence of glacial action is in the west of the peninsula, in the Dinaric mountains, and particularly in the northeast of Montenegro, where some 3,000 square

[1] 'Morphologische und glaciale Studien aus Bosnien, Hercegovina und Montenegro', I and II, *Abhandlungen der Geographische Gesellschaft in Wien*, 1890.
[2] 'L'époque glaciaire dans la péninsule des Balkans', *Annls Géogr.*, 9, 1900, 359–72. See also *Annls Géogr.*, 26, 1917, 189–208, 273–90.

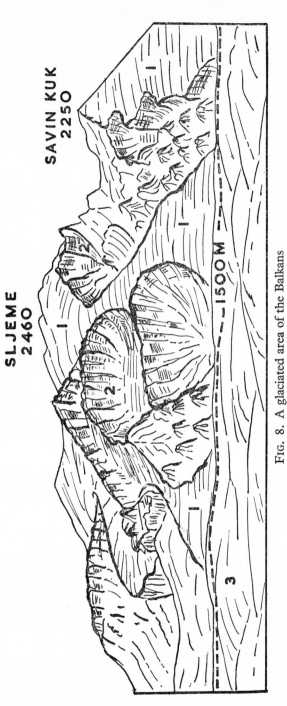

SLJEME
2460

SAVIN KUK
2250

1500 M

Fig. 8. A glaciated area of the Balkans

Redrawn from the original, this shows a number of cirques (2), the original mountain surface still surviving (1) and a peneplane surface at 1,500 metres (3). Originally published in *Annls Géogr.*, 26, 1917, 281.

kilometres were covered around peaks, Durmitor, 2,500 m., Rinja-
jevina, Žurian, Morǒcko Gradiste, all 2,300 m. A feature of special
interest was the relation of glaciation to the karst surfaces of the
peninsula, and on this subject Cvijić is informative and interesting.

Moraines in the Dinaric mountains have fresh vegetation throughout the summer in sharp contrast to the white dry karsts, and are
therefore valued by the transhumant farmers as pastures. Cvijić, in
noting the rich verdure and chalets on these moraines, used vegetational evidence as a key to morphology and even to geology like many
workers of his time. In the 1900 paper he also noted that some of
the karstic forms had been modified by ice, particularly in the
Dinaric mountains: for example the Ćaba glacier on the Treskavica
flowed into a karstic basin $1\frac{1}{2}$ km. long, which was deepened, polished
and striated. In addition, moraines held up lakes and formed a large
number of small dolines where the morainic material sank down
and glacial lakes found a subterranean outlet. The Durmitor area,
in the northeast of Montenegro, the most intensively glaciated area
of the Balkan peninsula (see above), provided the best evidence of
karstic glaciation as the massif has no normal valleys at all and ice
spread through a series of uvulas, normally aligned from northwest
to southeast, and was left in patches of dead ice during the recession
period. The forms of the uvulas were sharpened by ice action and
some of the dolines within the uvulas may have been initiated by
subglacial waters, as they appear to be far more numerous than in
uvulas not glaciated. But the main conclusion is that the karst forms
are preglacial in origin though in places modified by ice or glacial
waters: there is little evidence of postglacial karstic forms except
those formed in calcareous moraines, of which some collapsed and
developed a sub-surface drainage similar to that of limestone karst
itself. And, as noted above, some of the uvulas which had glaciers
have a very much larger number of little dolines than those not
glaciated, probably due to subglacial waters.

Cvijić was of the opinion that several of the mountains only came
within the snow limit by an uplift that began immediately before the
glacial period and continued into the glacial period. This view was
apparently first developed as a result of his researches in 1895 in
various mountain areas, including Olympus. He noted, for example,
that the Dinaric platforms had been laterally compressed and up-
lifted, and separated from the Metohija basin (drained by the upper

Drina and its tributaries) by an escarpment of 700–800 m. Similarly indicative was the fact that the Dinaric platform, and the glacifluvial terraces of the Zievna, the Mala Pike, the Morača and the Narenta fall—or indeed plunge—below the lake of Scutari or the Adriatic. In the littoral area of the Adriatic there are 'crypto-depressions', or basins whose floor is lower than the Adriatic. Cvijić argued that the existence of these crypto-depressions, with the plunging form of the glacifluvial terraces, was a proof of recent tectonic movements including some of postglacial date. Postglacial erosion was seen in the V-shaped valleys of the Metohija, 80–100 m. deep, becoming progressively shallower towards the crest and only reaching even 40 m. at some distance from the Adriatic and the Scutari lake.

It is beyond the purpose of this book to give a complete analysis of the work of Cvijić, but one must note, for example, that his long researches, particularly on karsts, bore widespread fruit. Much of the detailed material, which in time included many researches on caves, was published by the Serb Academy of Sciences and Arts in Belgrade. In a monograph published in French by the Serb Academy,[1] B. Ž. Milojević divided his work into four sections, the karst, with its relation to vegetation and life, the processes of karst formation, the variety of karst forms and finally, a world view of karsts, with examples from Jamaica, Great Britain, Northern France, Belgium, Germany and Moravia. Cvijić's work included a vast and searching analysis of karst forms, but the main conclusions can be stated in a few sentences. Having observed that streams may flow out at any point, above or below sea level, Cvijić proved that the main aquifer is the impermeable zone, which may be a non-calcareous bed below the karst or one made impermeable, perhaps only temporarily, by cementation. The smallest depressions are the dolines, and large features are uvulas, or depressions a mile or more long, which in time are united into poljes, much larger enclosed depressions. All may be affected in their origin and development by structural features: for example poljes may be associated with faults, or may be rift valleys (*graben*), or synclinal basins, even anticlinal basins. Gradually the idea was developed of a cycle of erosion in limestone regions, controlled at all stages not by sea level, but by the

[1] J. Cvijić, *La géographie des terraines calcaires*, Académie Serbe des Sciences et des Arts, monographie, tome CCCXLI, Classe des Sciences mathématiques et naturelles, Belgrade 1960. Preface by B. Ž. Milojević.

level of the subterranean water layer: at the end of the cycle all the limestone had disappeared except for some isolated residual blocks known as *hums*.

Cvijić as a human geographer

Had Cvijić done no more than contribute to physical geography, he would have acquired a reputation that most honest academics would envy. But by 1902 he was already showing a marked interest in the human geography of the Balkan peninsula, and the numerous papers written on this aspect formed the basis of his *Péninsule balkanique* published in 1918. Many critics might consider that academically his contributions to physical geography were of far greater merit and abiding value than the regional work: nevertheless, the regional work of 1918 and the articles preceding it were influential in re-shaping the map of Europe. True to his time Cvijić regarded *le milieu géographique* as of vast significance in shaping the lives of people, as this passage shows:

Geographical environment (*milieu*) influences not only the general march of history, the distribution of different civilisations, migrations, ethnographical distributions, the siting and the type of settlements and houses but also—directly or indirectly—the psychic character of populations. It is therefore an important aim of human geography to delineate the psychic character of populations in various natural regions, and to show the influence of geographical factors in the formation of these psychic characteristics. But human geography must not neglect other influences—historical, ethnic and social—which contribute to the formation of mental qualities, and are intermingled with geographical causes.[1]

Immediately following this statement, Cvijić notes the dangers of a subjective approach for any observer of mankind may be dominated by the spirit, the passions and the prejudices of his time that impartiality is hard to achieve. 'All depends,' he says, 'on the observer's qualities, on his accuracy of judgment' (*justesse d'esprit*). Whether Cvijić managed to be as impartial as he may have wished or not, he was certainly far more detached in judgment than many writers of his time, and his propaganda for Yugoslavia was so effective because it never appeared to be propaganda at all.

Ethnic considerations were regarded as supremely important at

---

[1] The influences of environment are discussed on pp. 81–5 of *La péninsule balkanique. Géographie humaine*, Paris 1918.

the Versailles conference and various maps showing the distribution of 'ethnic' groups were treated with considerable respect. But what exactly was meant by an ethnic group? Cvijić based his argument largely on the 'psychic' character of the population distributed through the Balkan peninsula, though he also observed the various racial types met in various areas: on this, however, he says comparatively little as in the Balkans such study was not advanced. Rather was he interested in costumes, dialects, views on life (*conceptions populaires sur le sens et la valeur de la vie humaine*), buildings, the daily routine of work, folklore, religious allegiance, and much more that in modern times would probably be studied by sociologists and social anthropologists. During the troublesome period of Balkan history in which Cvijić worked, it was hardly possible to rely on statistical sources, and it is clear that his main evidence was acquired during the annual field work, of at least a month's duration, carried out from 1888 to 1915. Modern geomorphologists who despise other forms of specialism, conspicuously regional geography, may here be invited to pause and ask themselves if Cvijić's way was not a good one, especially as it has been followed, with differences perhaps in method, by other productive continental geographers.

The first publication of direct human interest came in 1902, when Cvijić was thirty-seven years old. The Royal Serbian Academy began to publish (in Serbo-Croat) various observations made on human geography in the Balkan peninsula by Cvijić and, in time, his research students.[1] The first of these was in effect an atlas of the Dragečevo area (43° 40' N, 20° 20' E) with plates showing the origins of the families, their house types (with photographs and drawings), plans of houses and storage huts. A similar atlas volume of 1903 dealt with an area south of Belgrade populated by immigrants, and included plans of the villages, the use of the land for crops and a survey of buildings with some interior plans. A large number of photographs were included. Year by year similar material was published but gradually the scope of the work was extended, significantly perhaps in 1909, when the atlas volume published by the

---

[1] Serbo-Croat title, translated as *Problems of Human Geography in the Balkan Peninsula*, vol. 1, 1902; 2, 1903; 3, 1905; 5 and 6, 1909. These volumes are available at the Royal Geographical Society, London, but no. 4 is missing.

Serbian Academy showed the types of village houses in Hercegovina, with two maps of which one shows the agricultural areas and the summer pastures occupied by transhumant populations. In these earlier publications on the human geography of the Balkans one can recognize the interest in house types, forms of settlement, agriculture beside the villages and in the mountains, and above all the people, especially those known to have moved from other areas during the troublesome course of history in the peninsula. In essence, it was field work, with the author's own photographs and sketches.

By 1906, Cvijić was entering the controversial field of political geography by the publication (in Serbo-Croat) of his work on the geography and geology of Macedonia and ancient Serbia.[1] This was regarded by Vlahović in an obituary[2] as 'the first great fruit of his wanderings through the districts under Turkish rule'. This writer speaks of Cvijić as 'the first student of our people and the first scientist who went in person among the people themselves', for previously they had hardly been studied at all. Part of the 1906 work on Macedonia was translated into English (and also into French, German and Russian) as 'Remarks on the ethnography of the Macedonian Slavs'. The people were of varied origin, and Slavs, Greeks, Turks, Bulgars, Albanians and others were in a state of fusion. Each of the Balkan states regarded Macedonia as theirs, no doubt partly because it was traversed by the route to the port of Thessalonika, but the population had no sense of nationality, or inherent loyalty to Greece, Serbia, Bulgaria or Turkey. The Bulgar as well as the Serb languages were used but, Cvijić argues, the people of Macedonia were not ethnically Bulgar, though there was a Church of Bulgar origin (Greek Orthodox) which provided schools. The maps provided by the Germans, French and British showing Macedonia as Bulgar were false ethnographically and linguistically and the statistical data provided, particularly by the Turks, was untrue. In short, Macedonia was one of the most mixed areas of Europe, but by no means unwanted by the neighbours.

[1] The full title, translated, of this work (published in Belgrade) was *Sketch of the geography and geology of Macedonia and ancient Serbia with observations on southern Bulgaria, Thracia, the neighbouring regions of Asia Minor, Thessaly, Epyrus and Northern Albania.*

[2] Vlahović, M. S., 'Jovan Cvijić—scientist and patriot', *Anglo-Yugoslav Review*, 1, no. 3–4, 1936, 55–9.

A paper published in 1909[1] shows a further stage in the argument of Cvijić for a new Serbo-Croat state. Though it was clear that Turkish rule in the Balkans was neither desired nor likely to return, Cvijić pointed out that Serbia was in a favourable position in Turkey, having economic access to a considerable territory, but less fortunate when independent. The Dual Monarchy of Austria-Hungary was able to stifle her trade, particularly after Bosnia with Hercegovina was annexed in 1878. Having no direct access to the sea, Serbia could only use the Danube, where her boats were heavily taxed, or the difficult Morava-Vardar route. The argument was made that a Slav state should comprise Serbia, Bosnia-Hercegovina, Montenegro and the Sandjak of Novi-Bajar. Germany had encouraged Austria-Hungary to annex Bosnia-Hercegovina where, Cvijić alleged, the land problem had not been solved and the religious situation, between Roman Catholics and Greek Orthodox, was deteriorating into increasing bitterness. If a new state was made on the lines suggested, it would be 'an indivisible whole', for in Bosnia-Hercegovina was the very heart of the Serb community. The argument for the creation of this new state was based mainly on common nationality, by which 10 million Serbo-Croats and $1\frac{1}{2}$ million Slovenes would be united, and an ethnographic map was appended with an overprint showing the areas having more than 50 per cent Roman Catholics. It was also argued that a route to the coast was necessary, and this could only be acquired by adding at least part, if not the whole, of Bosnia-Hercegovina to Serbia.

Steadily through the early years of this century, the interest of Cvijić in human geography was growing. Eventually he explained that the ethnographical researches had been carefully planned on scientific lines. The field worker kept strictly to a line, as in a geological traverse, and observed the distinctive features of each district, with the inevitable result that differences which seemed to be gradual, so gradual as to be almost imperceptible, proved to be considerable in a long traverse. The word *psychique* frequently appears in Cvijić's writings, and he went deeply into the sociological aspects of the communities he studied. Although, as noted on p. 75, Cvijić almost apologized for his *Péninsule* when he said that during the Balkan wars of 1912–15 he was forced to consider problems of political and human geography, such problems had interested him long before

[1] *L'annexion de la Bosnie et la question serbe*, Paris 1909.

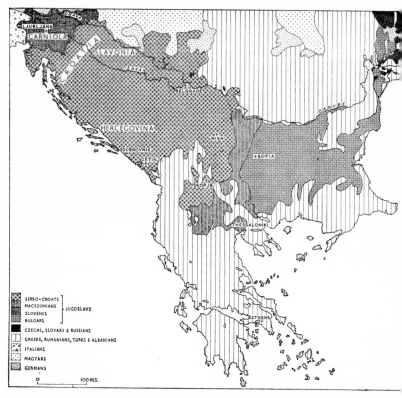

FIG. 9. Nationalities in the Balkan peninsula

Maps of this type were of considerable political significance at the time. Redrawn from *La péninsule balkanique*, 1918, endpaper map.

this time. It may be that, like many authors, he found the minor interest of many years developing into a major concern in time, and the classic work on the Balkans therefore came into existence relatively quickly, and almost through chance circumstances to the surprise of the author. *La péninsule balkanique* was largely the result of a series of lectures given in Paris from January 1917 to the end of the 1917–18 session.

The first task in *La péninsule balkanique* was to find a term to replace 'Turkey in Europe'. The term Balkan peninsula was already in common usage and regarded as reasonable, even though the Balkan mountains do not cover the entire peninsula, and the German term, Südost-europäische Halbinsel, southeast European peninsula, had political overtones, not unlike those given by German geographers to the 'regional' term Mitteleuropa. Nevertheless efforts to make a satisfactory northern boundary for the Balkan peninsula also had political implications. In the northwest, Cvijić found a suitable boundary in the meeting of the Dinaric mountains with the Alps, and included the Ljubljana basin as part of the Balkans, on the basis of physical affinity. But he encountered difficulties in making any satisfactory limits elsewhere on the north, for though the Sava river and the Danube to the Iron Gate might seem to be an appropriate natural limit, this did not coincide with the ethnic frontier, for on the north bank of these rivers, in Croatia, Slovenia, Syrmie, Bačka and Banat, the Serbo-Croats were a majority or at least an important section of the population. Though the argument for a unitary Slav state was not based on physical geography alone, physical geography could provide a useful line of argument.

As a whole, the Balkans had three main characteristics. The Balkan peninsula is 'Eurasiatic', showing both European and Asiatic influences: secondly, its physical features have permitted some union and penetration of external influences and thirdly, as a complement or even a contrast, there are areas of natural isolation. The first line of thought is based partly on physical connections, such as the merging of the Dinaric chains with the Alps, the clarity of the Carpatho-Balkan arc and the continuation southwards of the mountains into islands of the Aegean. Especially significant is the passage of Greek civilization into the Balkans and the development of Byzantine Christianity within the Balkans. The second line of argument follows from the first. A wide door was open for the incoming of continental

influences on the north, and the Danube tributary, the Sava, gave
an easy route leading to low mountain passes connecting the Balkans
with north Italy. The great routes through the peninsula followed
the Morava-Vardar and Morava-Marcia rivers; across the Dinaric
mountains transverse communication was not easy as it was necessary
to cross some broken and karstic ranges, but there were historic
routes inland from Cattaro, Dubrovnik (Ragusa) and Split. The
Adriatic coasts, with their numerous islands, had seen coastal movements from very early times. The third major characteristic was the
isolation and separation of certain areas, having *barrières ou de rem-
parts* which were obstacles both to climatic and biological influences
and to ethnic and other human movements. Deep gorges or thick
forests could be barriers at least as effective as high mountains, but
many of the mountain chains had several ranges and high plateaus
separated by deep enclosed valleys or closed karstic depressions:
even so, the Serbo-Croats had penetrated the Dinaric mountains.
The central Balkan mountains were in effect a human divide between the areas of patriarchal life to the north and of Byzantine
civilization to the south. There were in the Dinaric, Rhodope and
Pindus mountains 'zones of refuge', never thoroughly assimilated by
the Serbo-Croats. In such areas there were traces of the ancient
Balkan peoples who fled before the Slav invasion from the Adriatic
coast and from the Sava and the Danube. From the fourteenth and
fifteenth centuries the Serbs fled before the Turks, who never
effectively penetrated the remoter areas. Perhaps the best example
of a distinctive community in its own mountain fortress was Montenegro, almost completely isolated and virtually a social island, still
having a strongly entrenched patriarchal system. Similarly there
were areas in northern Albania and Hercegovina where a patriarchal
régime had become entrenched over several centuries. Of other
groups, the Albanians, the old Balkan population, isolated from the
great routes to the east by mountains, were living on the Adriatic
littoral, largely marshy in character, having no port but Valona and
not (at least until the 1930s with the occupation on Good Friday
1939) regarded as attractive by the Italians.

From his studies of the mountain peoples, Cvijić moves to a
regionalization of the Balkans which has clear affinities with the
work of Vidal de la Blache. The aim was to recognize the *pays* (*zupa*)
and group them into natural regions: the upper Morava, for ex-

ample, north of Nish is a fertile basin of considerable extent, but in most of the Morava valley, as also in the Vardar, there are numerous basins, 'like a chess board'. The distinction of many of the *pays* was accentuated by natural physical barriers, including forests, or by social barriers such as patriarchal areas between one *pays* and another. The grouping of the individual *pays* into natural regions rested largely on physical features, taken by Cvijić to include the morphological character, with the climate and vegetation inevitably influenced by the physical features. But a region, however well defined as a morphological unit, is not at unity in itself unless made so by human activity and historical developments. Cvijić maintained that while many of the *pays* had distinct ethnic and social characteristics, the major natural regions did not of necessity correspond to great historical or social units: for example the Marcia basin has a population of Bulgars, Turks and Greeks heavily influenced by migration movements, though by contrast there was more evidence of uniformity in the Morava-Vardar valley. On the Aegean region, Cvijić wrote with eloquence. The gulfs and lowlands are surrounded by mountains so difficult to cross that the region turns its back on continental Europe and looks outward across the sea to Asia Minor. All the truly Hellenic area has a pure Mediterranean climate but in Thrace there is a climatic gradient between Mediterranean and continental influences and here too great valleys open up ways to the interior so that both in climate and in landforms there is a transition from the Aegean to the Thraco-Mediterranean region.

It is not necessary here to analyse the whole regionalization of the Balkan peninsula, but enough has been said to indicate the methods used. Cvijić was strongly aware of the migration movements that, throughout history, had been characteristic of the Balkans, and he coined the term *metanastasique*, from the Greek *metanastis*, change of habitat, for the movement. Maps in the *Péninsule* show the course of these migrations, and full attention is given also to transhumance. One of the most fascinating maps of the *Péninsule* shows *zones de civilisation*, under the main headings of patriarchal, Balkan or modified Byzantine, Mediterranean (Italian), Turkish-Eastern. To these are added three more, as overprints, central European, western European (in towns) and finally—most significantly—the Serb national civilization derived from both central and eastern European influences. Though Cvijić showed remarkable tolerance for one of

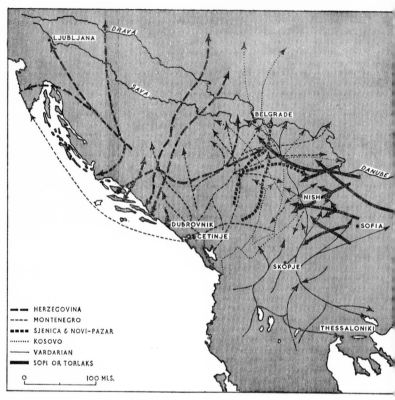

FIG. 10. Migration movements shown by Cvijić

The original map is far more complicated and printed in several colours: a selection of the various movements (*metanastasique*) is shown here. Selected and redrawn from *La péninsule balkanique*, 1918, endpaper map.

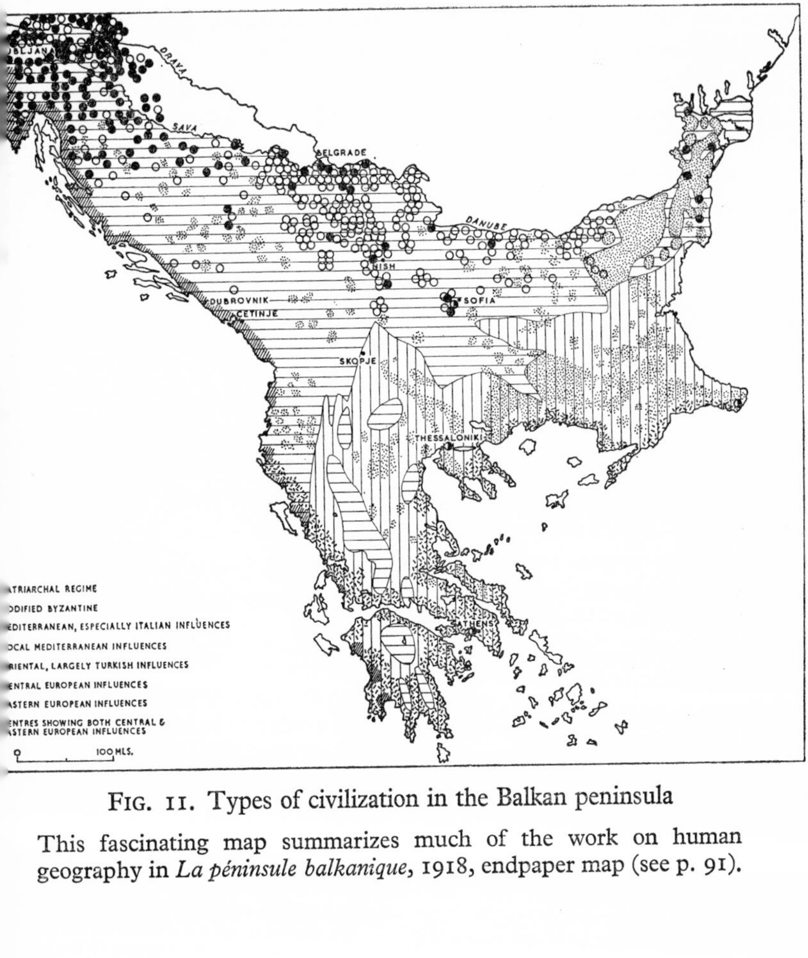
DRAVA
SAVA
BELGRADE
DANUBE
NISH
SOFIA
DUBROVNIK
CETINJE
SKOPJE
THESSALONIKI
ATHENS
REGIME
BYZANTINE
ESPECIALLY ITALIAN INFLUENCES
MEDITERRANEAN INFLUENCES
LARGELY TURKISH INFLUENCES
EUROPEAN INFLUENCES
EUROPEAN INFLUENCES
SHOWING BOTH CENTRAL &
EUROPEAN INFLUENCES
0
100 MLS.

FIG. 11. Types of civilization in the Balkan peninsula

This fascinating map summarizes much of the work on human geography in *La péninsule balkanique*, 1918, endpaper map (see p. 91).

his day and age, his main affection was shown to his own people, the Serbs. He writes with great warmth of the Serbs living north of the Danube who had preserved their own language and the Greek Orthodox faith. And as the power of Turkey faded, the main outside adversary was Austria-Hungary, which in Cvijić's view had the fatally wrong idea that the people should serve the state rather than the state the people. Cvijić saw the Austro-Hungarian state as the ruling dynasty with the army, the bureaucracy and the church as its instruments, clearly anxious to encourage the material development of the people but keeping them under close surveillance. The inclusion of the Croats with a new Slav state was based on the opposition of the Croats to the control of the Dual Monarchy, for in religion they were primarily Roman Catholic, and there were signs that the Roman Church, western in culture, had been gaining members from the eastern-orientated Greek Orthodox Church, notably in Carniola and Styria. In many cases this religious assimilation resulted from the marriage of immigrants with natives. In some areas certain families were venerated for their long ancestry, even though they might be in poor circumstances, particularly in the patriarchal societies of Montenegro.

Problems of the influence of environment on man, singly or in communities, were as fascinating to Cvijić as to many other geographers. He regarded this influence as direct, indirect and social in a wide sense. The direct influences were due to the terrain, climate and weather, all of which could affect people without any social intermediary. Thought and outlook might owe at least something to the existence of mountains with their summits, the wide plateaus with illimitable views, the tranquil or agitated sea, the karst with its mysterious grottoes, crevices, hollows and intermittent streams, the vast fertile plains that gladden the peasant's heart. Areas closed to the world outside, such as deep valleys, might breed an attitude of isolation, though very different impressions were given by a landscape of attractive hills intersected by green and laughing gentle valleys. Climatic influence was held to be direct, on the muscles, heart and lungs and—less easily assessed—on mental qualities. Man was strongly influenced by the transparency of a pure atmosphere, with light clouds, by sunrise and sunset, by the form and colour of snow, the changing form of rivers, lakes and seas, or of streams meandering through fields with crops. Similarly man is influenced by the varied aspects of the seasons, the richness and variety of flowers, the form of trees rich in foliage rising in a regular and firm pattern to the sky or struggling with twisted and stunted branches in the areas of Mediterranean climate. But climatic influences are so varied from one person to another that it is hard to assess them. And some aspects of nature trouble man, while others give him happiness and even lead him to finer thoughts: here too the influence must vary from one person to another, though without doubt man's imagination is affected by earthquakes and volcanoes, storms and tempests with snow or torrential rain, frosts and floods.

Environmental influence may come to man through social forces indirectly, and directly through economic circumstances such as the richness of the soil and its drainage and the associated crops and nutritive plants, pastures and forests. In some areas the produce of the sea, lakes and rivers and their use for navigation may affect life closely, even profitably. The availability of fuel and building materials may regulate even the details of life under a fairly simple economic régime with little outside trade. The third major influence, in essence social, was the movement and relation of people leading either to union and penetration or to isolation and separation. Population movements may transform the sentiments, thoughts and actions of national groups. Cvijić's insistence on the importance of migration movements is in line with that made by many geographers of his time, and some of a later age.[1]

The work of Cvijić

Cvijić had at least a share in the addition of Yugoslavia to the map of Europe: that he was in contact with Vidal de la Blache and his son-in-law, Emmanuel de Martonne, both of whom were shrewd political geographers, during the latter part of the 1914–18 war, is certain. His work on human geography bears a close relation to that of Vidal de la Blache, and is more thoroughly human in its concern with people than that of Jean Brunhes. Much that he wrote on the characteristics of the Slav people would now be regarded as amateurish sociology though it was new and interesting material for his readers at the time. And for all its apparent naïvety, it may give something nearer the spirit of the people than the masses of statistics that are sometimes provided in modern sociological works

[1] Cf. pp. 119–20.

where, in fact, the extracts from conversations still included by their writers often give the key to a situation. In the Balkans, Cvijić had a subject that he knew better than anyone else, which he had studied initially for its physical geography but with an observant eye for its people and their way of life. Unlike many later geographers, he had no particular need to consider industry, as there was very little: rather he was dealing with a rural community, with a peasantry. He shows the contemporary fashion in the strong attention given to historical aspects, and saw the effect of the past history in the existing distribution of peoples with varying ways of living from one area to another. In his analysis of the Balkans, he begins with the physical factors but goes on to show that these alone could not provide an adequate regionalization in areas diverse in culture due to historical circumstances. Like the geographers of his time, he was deeply conscious of environmental influences on people, but he was in no way deterministic and when he wrote the word 'environmentalism' had not acquired its present suspect association with determinism. It is easier to see the direct environmental influence among peasant people such as those he studied, than among more sophisticated urban and modern agricultural communities furnished with the resources of chemical fertilization and agricultural machinery.

No doubt, techniques of investigation have improved greatly since the days of Cvijić and much that he wrote may now seem impressionistic and even journalistic. Yet he used the statistics that were available and reasonably reliable, he was adventurous in his mapping (for example of religious groupings and of types of civilization), he was at once forceful and reserved in his treatment of environmental influence on people. The plain fact is that he was convincing, and his work appeared at the exact moment when it was needed to supply an argument for the new Yugoslav state. It was fortunate that there was some desire for (or at least guarded acceptance of) the idea of unity between the Serbs and the Croats with the Slovenes, for without this unity access to the Mediterranean, so badly needed by Serbia, would not have been possible. The post-1919 history of Yugoslavia showed that internal unity was not an easy growth, but rather an aspiration. And the new state naturally had enemies in Italy, deprived of much hoped-for territory on the Dalmatian coast, as well as of the dismembered Austria and Hungary on the north, not to mention Bulgaria, on the losing side in the

Balkan wars and in the 1914–18 war. Even so, what Cvijić helped to devise, in Yugoslavia, still exists much as he conceived it.

None of the obituaries says much of Cvijić, who died on 16 January 1927 at Belgrade, as a man. H. W. V. Temperley's work on Versailles[1] mentions his learning and shrewdness in the treatment of Yugoslavia's claim to self-determination, the guiding propaganda principle of the conference in 1919. No doubt Cvijić and his colleagues could point with pride to Serbia, inevitably the core of Yugoslavia, as the chief upholder of the doctrine of right against might and liberty against force in their resistance to Germany and Austria. But this again gives no clue to the personality of Cvijić, though from his writings—generally revealing of a man—it would appear that he was of a warm and sympathetic disposition. His claim that he became a human geographer through force of circumstances, even with reluctance, cannot remove the impression that he enjoyed writing on the human side of geography, for it is hard to conceive that any man could have written as much, with such verve and vitality, if he had not enjoyed it. But in the later years of a life of no great length—he was 61 when he died—he turned again to the physical aspects of geography in which, as shown earlier in this chapter, his contributions to glaciology and to the study of limestone landscapes, as well as others less original but nevertheless useful, would be enough to give him respect and even fame for many generations.

BIBLIOGRAPHICAL REFERENCES

Obituaries

Obituaries include those in *Geogrl Rev.*, 17, 1927, 240; *Annls Géogr.*, 36, 1927, 181–3, by L. Gallois; *Petermanns Mitt.*, 73, 1927, 102–3, by F. Machatschek; *Revue Mensuelle Géographique*, Cracow 5, 1927, 81–3 (in Polish); *Czasopismo Geograficzne*, Lwów, 5, 1927, 49–57 (in Polish, with a bibliography); *Wiadsmości Geograficzne*, Cracow 5, 1927, 17–20, 81–3. Later works include Milojević, B. Ž., 'Twentieth anniversary of the death of Jovan Cvijić', *Bulletin de la Société Serbe de Géographie*, 27, 1947, 49–53, with English summary showing that the work is mainly adulatory, and in Jovanović, P. S., 'Jovan Cvijić et la portée de son œuvre', *ibid.*, 29, 1949, 69–76 which deals with Cvijić's research, its fruits and mentions also the offices he held as Rector of Belgrade University, President of the Serb Geographical Society and President

[1] See also p. 74.

of the Serb Academy of Sciences. The interesting contribution of
M. S. Vlahović is noted on p. 86. The presentation volume of 1924,
Recueil de travaux offert à M. Jovan Cvijić, Belgrade, includes an
article in Czech by Daneš with an English summary and a remarkable
map of his annual journeys from 1888 onwards. Cvijić obviously kept
well-ordered diaries and field notebooks. This paper was re-published
with additional material as no. 46 of *Publications de la Faculté des
Sciences de l'Université Charles*, Prague 1925. It includes a photograph, but of a more conventional studies type than the drawing reproduced in this book. Prof. Dr. V. Hauffer of the Charles University,
Prague, kindly gave me a copy of this paper and of the photograph.

Works by Cvijić

1. *General physical geography and geology*

This interest is shown in 'Die Tektonik der Balkanhalbinsel mit
besonderer Berücksichtigung der neueren Fortschritte in der Kenntnis
der Geologie von Bulgarien, Serbien und Makedonien', *Comptes
Rendus IX Congrès géologique internationale de Vienne*, 1903. The more
southerly areas of the Balkans are treated in 'Grundlinien der Geographie und Geologie von Mazedonien und Altserbien: nebst Beobachtungen in Thrazien, Thessalien, Epirus und Nordalbanien', I
Theil, *Petermanns Mittheilungen, Ergänzungschaft* 162, Gotha 1908.
Another publication followed in 'Bildung und Dislozierung der dinar-
ischen Rumpffläche', *Petermanns Mitt.*, 55, 1909, 121–7, 156–63,
177–81. From 1906–11, material appeared in Serbo-Croat on the
geography and geology of Macedonia and ancient Serbia.

2. *Glaciation*

Of the papers, the first is 'Das Rila-Gebirge und seine ehemalige
Vergletscherung', *Zeitschrift der Gesellschaft für Erdkunde zu Berlin*,
33, 1898, 201–53: see *Annls Géogr.*, 26, 1917, 191. A wider area is
studied in 'Morphologische und glaciale Studien aus Bosnien, Herze-
govina und Montenegro', I and II, *Abhandlungen der Geographische
Gesellschaft in Wien*, 1890. Then followed 'L'époque glaciaire dans la
peninsule des Balkans', *Annls Géogr.*, 9, 1900, 359–72, and 'Neue
Engebrisse über die Eiszeit auf der Balkanhalbinsel', *Mitteilungen der
Geographischer Gesellschaft in Wien*, 47, 104, 149–95. A thorough
treatment is given in 'Beobachtungen über die Eiszeit auf der Balkanhalbinsel in der Südkarpathen und auf dem mysischen Olymp.',
Zeitschrift der Gletscherkunde, 3, 1–35, Berlin 1908: see *Annls Géogr.*,
26, 1917, 191.

3. *Limestone topography*

'Das Karstphänomen', in Penck's *Geographische Abhandlungen*,
Band 5, Heft 3, Vienna 1893, was the earliest of a long series of papers,
of which many were local studies published by the Académie Serbe
des Sciences et des Arts, Belgrade, and in the *Bulletin de la Société de*

Spéléologie, Paris. The theme of karst erosion was raised, 'Les crypto-dépressions de l'Europe', *La Géographie*, 5, 1902, 247–54, cf. *Annls Géogr.*, 26, 1917, 289. A local study with much general information, sections and two excellent block diagrams, is *Phases d'évolution du Karst de Moravie* (French summary, but text in Russian, Serbo-Croat), *L'Académie des Sciences*, 108, Belgrade 1923. Important principles are discussed in 'Hydrographie souterraine et évolution morphologique du Karst', in *Recueil des Travaux de l'Institut de Géographie Alpine*, Grenoble, tome 6ème, 1918, 375–426. The theory that river captures could occur underground is stressed in *Circulation des eaux et erosion karstique*, Zagreb 1925. A posthumous work, with a short preface by B. Ž. Milojević, is 'La géographie des terrains calcaires', *Académie serbe des sciences et des arts*, vol. 341, Belgrade, 1960: the manuscript was revised by E. de Martonne. A large and excellently illustrated text in two volumes notable for the long section on karsts, is *Morphologie Terrestre*, Belgrade 1924 and 1926.

4. *Atlases*

The first was the geological atlas of Macedonia and old Serbia, Belgrade 1903, in Serbo-Croat, but with keys to the maps in French. Two later ones are *Geologische Skizze zur Entwicklungsgeschichte des eisener Tores*, Gotha 1908, and *Entwicklungsgeschichtliche Karte des eisener Tores*, Gotha 1908.

5. *Human and regional geography*

In Serbo-Croat, published by the Serbian Academy of Sciences, there are several volumes on population from 1902, including some student theses. From 1902–9, atlas volumes were published on problems of human geography in the Balkan peninsula, with a wide range of maps, drawings and photographs: the text is in Serbo-Croat with captions in English and French. A paper of 1906, *Remarks on the ethnography of the Macedonian Slavs*, was translated by Anne O'Brien: it was also translated into French, German and Russian. The progress of propaganda is seen in *L'annexion de la Bosnia et la question serbe*, Paris 1909, with strong arguments for the creation of a new state. A suggested boundary is given in a paper entirely in Serbo-Croat, but with a French title, *L'unité et les caractères ethnopsychiques des Jugoslavs*, Nish 1915. Several papers were collected together in *Questions balkaniques*, Paris and Neuchâtel, 1916: the first essay is a discussion of M. I. Newbigin's *Geographical aspects of Balkan problems*, London 1915, in its day an interesting and much-studied work. Summarizing much of the continuing work on population movements, 'Les movements metanastasiques dans la Péninsule des Balkans' appeared in *Monde Slave*, 1, Paris 1917: much of this material appears in *La péninsule balkanique*, Paris 1918. Further papers include 'Unité ethnique et nationale des yugoslaves', *Scientia*, 23, Bologna 1918, lxxiv–6 and 455–63; 'Frontière septentrionale des yugoslaves', Paris 1919; 'Des

migrations dans les pays yugoslaves: l'adaption au milieu', *Revue des Études slaves*, 3, Paris 1923. The Serbian Royal Academy published (Serbo-Croat) in its ethnographical magazine, 1922, a more detailed map of the settlement and origin of the Balkan peoples than that in *La péninsule balkanique*.

CHAPTER 5

Ellsworth Huntington

A STUDENT OF CIVILIZATION

'No nation has risen to the highest grade of civilization except in regions where the climatic stimulus is great. This statement sums up our entire hypothesis.'[1] Ellsworth Huntington certainly knew his own mind and knew what he wanted to say. He wrote voluminously and even though there is a certain measure of repetition between one work and another, considerable industry is represented by an output of twenty-nine volumes, chapters in twenty-seven other books and over 180 articles. True, some of the textbooks, of which a few are still widely read in new and revised editions,[2] were written in collaboration but for sheer vitality and pertinacity few geographers have rivalled Huntington. No obituary tells us if he had his home subjected to the permanently or intermittently stimulating changes of climate that he regarded as so desirable for human progress. Perhaps with undue modesty, another obituary says that he is 'one of the few American geographers of whom the whole world knows':[3] almost every student of geography has read one or more of his works, or has some idea of his main theme—the influence of climate on humanity—yet his work was broader than many people realize, and his mind had more critical judgment than the casual reader might expect. Like Hamlet, he is full of quotations, but the quotations give only part of the story and only a fractional suggestion of the great range of his thought and reading.

Huntington was born in a parsonage at Galesburg, Illinois, and throughout his work showed an interest in the mental qualities and outlook of people with an appreciation of their religious views that tempers his apparent preoccupation with their physical qualities.

---

[1] *Civilization and Climate*, New Haven 1915, 270.
[2] A complete bibliography is given at the end of the obituary by S. S. Visher, in *Ann. Ass. Am. Geogr.*, 38, 1948, 38–50.
[3] In *Geogrl Rev.*, 38, 1948, 153–5.

None of his travels gave him greater satisfaction than his time in Palestine, on which he wrote a book full of vivid regional description and deeply impregnated with a knowledge of the scriptures as well as of George Adam Smith's *Historical Geography of the Holy Land*.[1] The book on Palestine,[2] with his *West of the Pacific*, based on a tour of Japan, Korea, China, Java and Australia, were described by S. S. Visher[3] as 'gems of descriptive writing', and in many of his books there are descriptive passages that show he had been there: *West of the Pacific* is, however, merely a chatty work of travel. Geographically, he was a child of his time in aspiring to a world view of civilization, based on experience of a vast range of places and on wide reading. Like Semple, Vidal de la Blache and other 'human' geographers, he used the detailed accumulation of field research to establish principles—in his case, what virtually amounts to a pattern of human behaviour. He admitted in 1920[4] that when he heard or read a good idea he assimilated it, and later gave it out as if it were his own, not knowing where or when he received it: in fact his books are less completely documented than those of E. C. Semple or Vidal de la Blache. In later years, he became increasingly interested in eugenics, and read widely in biology, psychology, world history (notably Toynbee) and many more fields of study. Even so, he still held that climate was one key, but emphatically one key only, to human history, to the development of each nation and to human life everywhere, however air-conditioned the circumstances of life might be. More than once, he regarded modern American central heating as debilitating to people as it was both too warm and too dry to be healthy: moderation, though not necessarily the chilling draughts of English houses, was perhaps better. He notes that

. . . the uniform dryness within doors does almost untold harm in parching the mucous membranes and thus rendering us peculiarly liable to colds, grippe and similar ailments which often lead to serious diseases such as pneumonia and tuberculosis. . . . If the conditions inside our houses could be like those that prevail in the autumn, the general health of the community would be much improved.[5]

Born in September 1876, Huntington died in 1947, writing to the last. For most of his professional life, he was a Research Associate of

[1] First published 1894: over twenty later editions.   [2] See pp. 109–11.
[3] Op. cit., 40.   [4] *World Power and Evolution*, New Haven 1920, 9.
[5] *Civilization and Climate*, 1915, 289.

Yale University, free from routine teaching duties and—still more valuable—from administrative responsibilities, and his teaching was limited to short periods of service at various American universities, to the conduct of occasional seminars and the supervision of graduate theses. He was able at various times to spend long periods abroad, and his travels began immediately after his graduation in 1897, when he went to teach in Turkey, at the Euphrates College, Harput, where he stayed till 1901: he spent the summers in field work, notably on the canyons of the river Euphrates. Returning to America he studied for two years at Harvard, largely under W. M. Davis, then in the full flower of his vigorous middle age. Out of this fortunate association there came the recommendation that Huntington should join the expedition financed by the Carnegie Institution to Turkestan, under Mr Raphael Pumpelly. At this stage, Huntington was a research assistant of the Carnegie Institution, and his main task was to assist W. M. Davis in physiographic survey. From May 1903 to July of the following year, he worked mainly in Russian Turkestan, with one month in Chinese Turkestan and four months in east Persia during the winter. In 1905, he went with Robert Barrett on another expedition to Chinese Turkestan; having arrived in India in February, he travelled north to the Vale of Kashmir, crossed the Himalayas in May, worked on the Tarim and Turfan basins and ultimately returned home by Siberia and European Russia in May 1907. These expeditions laid the foundations of much of Huntington's career as a geographer, and two long sections in the report of the Pumpelly expedition include such titles as 'changes of climate' that were to become frequent and characteristic of his later works.[1] But the work that brought Huntington's name to a far wider public was the *Pulse of Asia*: readable, even dramatic, vivid and far more intellectually satisfying than most works of travel. Even now, it is well worth reading but at the time it must have enthralled those who discovered it. Here was a new and original light on the little-known history of Asia.

[1] *Explorations in Turkestan with an account of the basin of eastern Persia and Sistan, Expedition of 1903, under the direction of Raphael Pumpelly*, Carnegie Institution of Washington, Publication no. 26, 1905. The sections by Huntington, pp. 159–317, are 'A geologic and physiographic reconnaissance in central Turkestan', and 'The basin of eastern Persia and Sistan'.

Man in nature

Most widely known as a writer on climate and man, Huntington was a well-trained and observant student of geomorphology using his early work in geology as the background for his consideration of physical processes such as the formation of terraces showing the former levels of lakes. On this basis he worked out his theories of climatic oscillations that were to become the basis of his teaching and one of his main contributions to learning, for the idea that climate had changed significantly during postglacial times was by no means universally accepted fifty years ago. Huntington found himself led on from the physical to the human aspects of life, and begins his book on *The Pulse of Asia*[1] with the statement that the advance of the biological sciences, botany, zoology and physiography, during the previous fifty years was made possible by the work of Charles Darwin, whose theories of evolution had implied a relationship that made a mass of previously unrelated facts explicable. Just as the biological sciences formed a related group, so too the anthropological sciences, geography, anthropology, history and sociology, formed a group possessing a unity as great as that of the biological sciences. And in this group, geography 'tells us not only what forms of plants and animals live together in mutual dependence, but also why the human inhabitants of a given region possess certain habits, occupations, and mental and moral characteristics and why they have adopted a certain form of social organization'.[2]

This statement is confident in tone, and Huntington regarded the influence of environmental conditions as crucial in all human study. He goes on to say that man's relation to environment was inevitably clearer among more primitive societies, for among highly civilized peoples the environmental influence was obscured, apparently almost totally in some cases, by the mixture of races, man's control over nature, the rise of great religious or ethical ideas, racial hatreds and dominant personalities. Even so, the influence of physical conditions in man was strong, and it was the task of the geographer to bring it to light. Huntington was not content in his work to consider only primitive societies, like some modern anthropologists, but dealt also with sophisticated industrial societies, for he shared with

[1] *The Pulse of Asia* was published in 1907 in London, Boston and New York.   [2] Ibid., 2.

various contemporaries the desire to study the whole world of humanity, past and present, with an eye to the future. Equally he was concerned with the mental and moral condition of man as well as with his physical life.

Broad generalizations early became a feature of Huntington's work, and brought a number of rebukes from those of more cautious mind, such as J. Scott Keltie who wrote that he was 'one of the most active and original of our younger geographers, whose imagination may perhaps want a little of the taming that comes with years.'[1] But Huntington remained young and adventurous in mind. From the beginnings in *The Pulse of Asia* to his last work, many of his views remained constant. He formed a clear view that neither climate nor changes in climate had been adequately recognized as influences on man, and that climate explained both the nomadic and plundering propensities of the Arabs and the easy-going tilling of the soil by the Italians. In central Asia disorders, wars and migrations had been caused by climatic change, with a consequent mixing of races that resulted in new habits of life and alterations of character. Not only so, but the periods of desiccation in central Asia had impelled the outward movements of people as migrants and conquerors and so plunged Europe into the confusion and miseries of the Dark Ages. Possibly Huntington had a simpler view of history than the facts justify—for example, it is untrue that the Dark Ages were as dark as many suppose—but the attraction of his theories lay in the light they shed on many historical movements. The idea of cycles came to Huntington early, partly through his work in physical geography. Summarizing the story of central Asia, he wrote:

Many facts in western and central Asia suggest that during the past two or three thousand years the climate of those regions has passed through four successive phases. Up to the first or second century of our era, it appears to have been considerably colder and moister than at present. Then, for several hundred years, it grew rapidly warmer and drier until in the fifth and sixth centuries the desert regions were even more arid than today. During the succeeding medieval epoch, the climate again became slightly cooler and moister; while during modern times there is a general, though slight tendency toward aridity.[2]

[1] Keltie, J. Scott, 'Thirty years progress in geographical education', *Geogr. Teach.*, 7, 1913–14, 224.
[2] *The Pulse of Asia*, op. cit., 43–4.

In times of adequate rainfall, many areas of central Asia supported a large population of farmers and artisans, but as the rains diminished they gave way to a few wandering nomads.

Field work in central Asia was the basis of these views: during a long sojourn in the Tarim basin, Huntington noted that though there was only rainfall of an inch or two in the central area the mountains had 25–30 inches which made possible some agriculture on its margins or in the terraced valleys. But signs of the decay of vegetation were numerous: plants were drying up, the tamarisks were withering, the poplars dead or dying except in the moister places and beds of dead reeds covered several square miles. Ancient roads had been abandoned as no water was available beside them, and the springs once used by men or animals had disappeared. During many of his travels, Huntington marked the sites of abandoned villages: in the Tarim basin, historical evidence was given by the signs of waxing and waning of the lake, Lob Nor, during the past 2,000 years. At the beginning of the Christian era, it was comparatively large, perhaps 75 miles in each direction even though a good deal of its water was diverted to supply populous rural areas and towns. In the early centuries of the Christian era, particularly from the third to the sixth centuries, it decreased greatly in size due to aridity and many towns, rural areas and trade routes were abandoned. From the ninth to the sixteenth centuries there was an epoch of more favourable climate with an expansion of the lake and the founding of new towns trading with the re-occupied rural areas, but from late medieval times to the present aridity set in again, with the consequent withdrawal of population. Similar conditions were observed in the Turfan basin. Huntington spent several months among the Chantos, an agricultural people of the Tarim basin, and contrasts them in character with the nomadic Kirghiz of the Tian Shan mountains. He found the Kirghiz polite, brave, self-reliant, hardy, hospitable but lazy when the opportunity of being lazy arose: the Chantos, on the other hand, had less marked good or bad qualities:

Among the good qualities, the chief are gentleness, good temper, hospitality, courtesy, patience, contentment, religious tolerance and industry; among the bad are timidity, dishonesty, stupidity, provincialism, childishness, lack of initiative, lack of curiosity, indifference to the sufferings of others, and immorality. . . . Determination, courage, aggressiveness, insolence, undue curiosity, violence, fanaticism, and the like, are almost unknown among the Chantos. Neither their good nor their bad traits demand any great exertion of will or purpose. On the one hand, there is no public spirit; almost no one exerts himself for the good of the people as a whole. On the other hand, crimes of violence, and even theft, are very rare.[1]

Huntington regarded such settled agricultural people as somewhat negative in character, largely because their lives were comparatively easy, and in various places he spoke of mountaineers as possessing more sterling qualities as a response to their harder environment. There are in Huntington's work many signs of a somewhat austere judgment of character, perhaps associated with his Puritan background.

Having studied central Asia, Huntington proceeds to write nearly thirty pages at the end of *The Pulse of Asia* under the title, 'the geographic basis of history'. The first part of this section is in part a summary of what went before—incidentally, he did not spare his words and under modern conditions would be compelled to compress his writing—but new ideas are put forward which were developed later. Virtually permanent factors of the environment include relief, the distribution of water, and major differences of temperature, changeable factors include accidental events such as earthquakes, volcanic eruptions, changes in the course of rivers, hurricanes and the like, any of which may be dramatic and even devastating: vastly more important, however, are changes of climate of longer or shorter duration. These were attracting attention in various ways when Huntington wrote: using some material given by educational psychologists working in New York and Denver, he pointed out that in damp muggy weather people feel disagreeable, and suppose themselves ready to do all sorts of evil things . . . but do not do them. They merely have the relief of talking and feeling cross, apparently rather like Miss Pole in Mrs Gaskell's *Cranford*: ' . . . She took me so by surprise, I had nothing to say. I wish I had thought of something very sharp and sarcastic; I dare say I shall tonight.' But, Huntington goes on, in dry weather there is a surplus of energy, and in Denver it had been observed that in dry periods the amount of crime among adults and of misconduct among children increases

---

[1] *The Pulse of Asia*, 225. See also, on Kirghiz, 132, and the chapter on 'The waxing and waning of Lop Nor'. This name is now transliterated as Lob Nor.

enormously. Although people commonly mention their reaction to weather, it appears to differ considerably from one person to another.

Eduard Bruckner suggested that once in every thirty-five years the world passes through a climatic cycle, in which there are two extremes: at one the climate of continental regions for a series of years is unusually cool and rainy, with low barometric pressures and frequent storms, and at the other it is comparatively warm, sunny and dry in summer but cold and clear in winter. The differences between these extremes are greatest in continental interiors, and diminish towards coasts: possibly the colder periods are associated with times of sunspot formation. Huntington used his own evidence and that of others to urge that there had been great changes in climate, including some of long duration to be measured in centuries (v.s.) as well as others of short duration.[1] In Palestine and Egypt the area under cultivation had diminished over the preceding 3,000 years, and there was evidence of the desiccation of the Sahara. Julius Cæsar's books suggest that the climate of Europe was much colder in Roman times than now, and Gibbon in his *Decline and Fall of the Roman Empire* thought that this was due to the removal of forests. On perhaps rather more certain grounds, Huntington mentions three periods of the nineteenth century known to be dry: in the first of these, 1830–40, the Tarim basin had a severe drought in which some villages were abandoned and people migrated upstream. Persia had destructive famines in the 1830s, and also in the two later dry periods, 1865–75, and 1887–97. Turkey had disasters in all three periods: from 1829–33 all the Balkan states were in an uproar, there were rebellions in Asiatic Turkey, and wars with Egypt and Russia. Further troubles came in the 1860s, and in 1874 fresh disturbances led ultimately to war with Russia in 1877. And in the 1890s there were massacres in Armenia, of which vivid details are given: it was apparently the practice to tie the Armenians up in bitter winter cold and pour water over them to make them freeze. Even in the United States there were political crises in 1837, 1874 and 1893, which leads to the statement that 'The idea that financial crises and political changes in the United States may be genetically connected with famines and revolts in Asia suggests a hitherto unsuspected unity of history.'[2] The clear relation of famine to climatic change has long been appreciated, but Huntington also urged that

[1] See pp. 104–7.   [2] *The Pulse of Asia*, op. cit., 378.

deficiency of rain could cause insurrections, wars and massacres, partly through failure to pay taxes, and he underlines his point by descriptions of Turks beating up Kurds. But in some cases, drier conditions can be beneficial, as for example in Kashmir, where higher temperatures and diminished snowfalls made for progress.

The climatic theory develops

From the initial studies in Central Asia, the view of Huntington spread gradually outwards as foreshadowed in the last pages of *The Pulse of Asia*, which ends on a somewhat melodramatic note: 'With every throb of the climatic pulse which we have felt in Central Asia, the centre of civilization has moved this way or that. Each throb has sent pain and decay to the lands whose day was done, life and vigour to those whose day was yet to be.'[1] Immediately before this he puts forward his climatic hypothesis of history (so named), which is that man has always made the most rapid human progress under essentially similar climatic conditions: the summer must have a sufficient degree of warmth and of rainfall to make agriculture easy and profitable, but yet must not be enervating and the winters must be bracing though not frigidly deadening. With forethought, it should be possible for man to support himself on agricultural produce for the whole year, but he must be under the stimulating necessity of forethought, as no man is likely to develop his powers if he can live merely by picking up coconuts and planting a few vegetables now and then. From many passages in his books, it is clear that Huntington was at all times an admirer of action rather than of contemplation. Comparatively clear dry air with high barometric pressures appear to be favourable to human progress. Like Herbert Spencer and Carl Ritter, Huntington saw that man had become adapted to cooler and moister climates than those in which civilization first developed under the stimulus of irrigation, for example in the Tigris-Euphrates trough and riverine Egypt, but he believed that the physical changes of climate had been considerable. And he found many of his views confirmed, or at least supported, by a study of Palestine based on a visit from February to October 1909 as a member of the Yale expedition.

The book on Palestine[2] is of interest not only for its excellently

[1] Ibid., 385.
[2] *Palestine and its Transformation*, London, Boston and New York 1911.

detailed study of physical features and of climate, but also for its expression of the views on race and migration that were to attract considerable attention in time.[1] The major part of the book consists of a detailed study of the physical regions of Palestine enlivened by reminiscences that, if not invariably relevant, add to the readability. He shows that within the relatively small area of Palestine there are pronounced contrasts of scene and natural resources, which have persisted through centuries of climatic change. As Palestine lies on the border between the great desert tracts of Asia and the better-watered countries of the Mediterranean, the effects of increasing aridity have been marked. Unreliable as records may be, the drier parts of Palestine have many ruins of large cities now abandoned or occupied only by a few peasants, with relics of former villages and farmsteads, and old terraces and walls made during times of agricultural settlement. Forests were probably never extensive, but some trees could be cut for local use where none grow now. A third line of evidence is the withdrawal from ancient trade routes such as that through the north of the Sinai peninsula, which had more water along its course in the days of the Exodus. As with the Lob Nor, so Huntington investigated the Dead Sea, and found major strand lines above the present level at 1,430 ft, 540 ft, 340 ft, 300 ft and 250 ft: the highest of these may indicate some crustal movement, but the others show that more water was available so that the sea may have stretched some 30 miles south of its present limit. In the time of Christ, Palestine must have been a far more attractive land than it was nineteen centuries later and Galilee a paradise compared with what it is now,[2] for the decline could not, in Huntington's view, be ascribed to agricultural inefficiency: he showed that many peoples on the desert fringe farm with minute care and only lack rain to make their efforts fruitful.

Some of the evidence accumulated on Palestine is compared with that from the Caspian Sea, investigated during Huntington's earlier travels. Probably about 3000 B.C. the Dead Sea was high, and at this time Egypt and Babylon were prosperous, though they were eventually brought low by invaders from the interior of the Old World. From about 1100 B.C. or slightly later to the time of Christ, comparatively moist conditions were generally prevalent, and the period was marked by the success of Israel in Palestine, of Assyria

[1] See pp. 119–20.   [2] *Palestine and its Transformation*, op. cit., 303–36.

and Persia far to the east, of the Greeks in their islands and peninsulas and of the Roman civilizations in Italy and beyond it. But as
climatic changes came once more, progressively from about A.D.
325–360 to A.D. 550, a long series of invasions from the interior
heartland of the Old World brought renewed devastation. After the
seventh century, however, conditions improved, and in A.D. 920 the
Caspian was 29 ft above its present level, and by 1100 the Dead Sea
was higher than now. From approximately 1300 the aridity became
marked again, and in Palestine after about 1250 all forms of art and
architecture practically disappeared, except in the better-watered
areas that remained permanently habitable. As an indirect result of
climatic change, Huntington suggests that there may be modification
'in the character of a race'.

A thousand years of life under the bracing conditions of the Judean
plateau, as it was in ancient times, may have eliminated many weak
elements from the Hebrew race, and given it a strength far beyond
that of the present inhabitants, a strength commensurate with the
greatness of its contribution to history . . . the decline may be due to
disease, especially malaria.[1]

Such ideas were developed far more strongly later. In his book on
Palestine Huntington shows his skill at constructing an ending, in
this case one that is revealing of the author, but a quotation can
speak for itself:

When nature set the seal of doom upon the country, she touched the
physical. She blasted progress, drove men to anarchy, and despair,
and killed great thoughts and aspirations. Yet the ideas already evolved
would not die. Christ took them, ennobled them, and made them the
greatest of earthly forces. They were emancipated from material
control. They spread beyond the sphere of desolation, carrying with
them life and faith. They proved that little by little the mastery is
passing from the lower reaches of nature—the material—to the higher
realms—the ideal.[2]

It is quite untrue that Ellsworth Huntington regarded men as
puppets dancing to the stimulus of climate in spite of his very firm
statements on the relation of climate and civilization: indeed, his
increasing interest in heredity, eugenics and population movements
suggests that he did not wish his apparently rigid views to be
accepted without modification. In the academic circumstances of
his time, the argument for climatic change had to be pressed forward

[1] Ibid., 409–10.   [2] Ibid., 415–16.

strongly as many were of the opinion that changes had been slight or non-existent. In 1914, he published the results of his work on America,[1] which opens with comments on the work done by the various national weather bureaux and by researchers on the climates of geological times: between the contemporary and the geological material, an historical study was required, particularly for the preceding 2,000–3,000 years. Only by the study of past variations could one hope to understand those of the present and of the future, and a mathematical investigation of the chief effects of present climatic conditions could do much to show how far human habits, customs, psychological traits and mental character are influenced by physical environment. Huntington was anxious to find a valid scientific foundation for his views, and worked in association with the Department of Botanical Research of the Carnegie Institution at Washington. He spent four periods of field study on arid America, the first, in 1910, based on Tucson, Arizona, in the desert and semidesert landscape extending to the gulf of California, the next on the measurement of the growth of sequoias in the Sierra Nevada, the third on lakes and ancient rivers of South Mexico and Yucatan with more measurement of sequoias, and the last in Guatemala to see the relations of the rivers of the Maya civilization to the physical surroundings and existing vegetation. From these lines of enquiry, he reached the conclusion that in spite of certain disagreements both in America and Asia the general climatic history appeared to have been characterized by similar pulsations. In Guatemala and Yucatan the presence of magnificent ruins in the middle of dense forests suggested that the climate was formerly drier; apparently the forests of Yucatan and the surrounding regions had alternately increased and diminished in size, and these changes could be correlated with conditions in California, for the forest increases came when California was arid and the decreases in times when California was moist. Further possible correlations could be made with the rise and decay of salt lakes, on which some fine work had already been done by American geologists, and with the varying conditions in Mexico, of which Huntington writes from his own observations, noting especially the oscillations of the margins of agricultural settlement as people were tempted forward in wet years and repulsed in dry ones.

[1] *The Climatic Factor, as Illustrated in Arid America*, Washington D.C. 1914.

How then were the Maya people able to achieve so much in Yucatan? In *Civilization and Climate*[1] he suggests that the Mayas made their greatest progress when their country was blessed with a climate somewhat like that of north Italy, warm but having frequent cold waves from the north as one of its stimulating characteristics. This, however, was not the only explanation as Huntington shows in a vivid passage:

The normal decay of races, the interplay of historic forces, the invasion of barbarians, the decadence due to luxury, vice, and irreligion, the change of the centre of World power, each or all of these causes, or any others usually appealed to by historians, cannot explain the matter. The question is not why the Maya civilization rose, nor why it fell. We may assume that it arose because it is the nature of a young and vigorous race to make progress, and it fell because it is the nature of an old and exhausted civilization to decay. The assumption does not help us in the least, for it does not touch our problem. Today the most progressive and energetic people of Guatemala, its densest population, its greatest towns, its centre of wealth, learning and culture, so far as these things exist, are all located in the relatively open, healthful, easily accessible and easily tillable highlands; in the past these things were located in the most inaccessible, unhealthful, and untillable lowlands.[2]

Huntington was strongly of the opinion that archaeologists could help to solve many mysteries such as that of the history of the Maya civilization, and stated the evidence for the rise and fall of water in Mexico City, some of which he gathered from the work of Alexander von Humboldt (1769–1859), who is generally regarded as one of the fathers of modern geography:[3] surely it was not without significance that the climate of Mexico had passed through fluctuations such as those of Asia on the one hand, and of more northern areas of America, such as California and New Mexico, on the other.

In 1915,[4] the theories were spread further: and in the preface to the alluring-titled *Civilization and Climate* Huntington says that he first developed his theories of climatic pulsations during the Pumpelly expedition from 1903, which confirmed the view of earlier geographers such as Reclus and Kropotkin[5] that central Asia was

[1] *Civilization and Climate*, New Haven 1915, 267.
[2] *The Climatic Factor*, op. cit., 215.   [3] Ibid., 96.
[4] *Civilization and Climate*, New Haven 1915, 134.
[5] Jean Jacques Élisee Reclus, 1830–1905: Peter Alexander Kropotkin, 1842–1921.

formerly moister than now. His subsequent travels, discussed above, confirmed this thesis, and he therefore accepted the view of Albrecht Penck that there had been a shifting of the earth's climatic zones alternately toward and away from the equator. This explained so much: for 1,000 years western and central Europe had been the great centre of civilization, partly because it is the only large region of the world that for many centuries has enjoyed a highly stimulating climate. But that is not all, for 'the position of western Europe, the strong racial inheritance of its people, their legacy from the civilizations of the past, their inspiring religion, their political freedom, and their many powerful institutions, all play an essential part'.[1] On the other hand, climates can be too stimulating as on the coast of California where people, Huntington says, indulge in a senseless round of activity that induces nervous disorders and a high number of suicides, especially in San Francisco.[2] Huntington uses here an analogy between the Californians and horses so beaten and urged to the limit of endurance that they break down: the analogy with horses is used several times in his books. Apparently the American climate is not entirely favourable to the growth of civilization. In the South there is far less energy, vitality and education than in the North, with fewer men who rise to eminence than in the North not through any deficiency of innate ability but because of the adverse climate. The North has a wonderfully stimulating climate for much of the year, but the people

suffer sudden checks because of the extremes of temperature. These conditions favour nervousness, and worst of all they frequently stimulate harmful activities. That, perhaps, is why American children are so rude and boisterous, or why so staid a city as Boston has six times as many murders as London in proportion to the population. Our country takes immigrants of every mental calibre, and then stimulates some to noble deeds and others to commit murder, break down the respect for law, and give us city governments that shame us in the eyes of the world. All these things would apparently not happen to such an extent were our climate less bracing and did not its extremes often weaken the power of self-control.[3]

Well, that's what he said. He also said that the Germans suffered from excessive stimulus.

What then are the ideal conditions for humanity? Huntington

[1] *Civilization and Climate*, op. cit., 253.   [2] Ibid., 134.
[3] Ibid., 286.

obviously thought for years about this, and in some of his earlier works was feeling the way to his ultimate conclusions:[1] it is therefore fairer to quote from one of his later works. The best physical health is attained with an average temperature of 64° F., ranging from 55° F. at night to 70° F. by day, but for mental health an outdoor temperature averaging 40° F., with mild night frosts and about 50° F. in the daytime, is most favourable. This conclusion was based on the work of 1,000 students at West Point and Annapolis, but a university teacher might be tempted to infer that if the weather is fine, students will go out and play, but if it is cold they will work: at least some will. For factory work a mean temperature of 60° F., slightly lower than that for the best physical health, is most desirable. Frequent changes of temperature and an average humidity of 80 per cent are favourable. Huntington thought that American homes were overheated in winter,[2] as the internal temperature was 65°–75° and the air dry, so that they had the climate of the interior of the United States in summer, which he correlated with a death rate 10 per cent above the average. It would therefore be sensible to vary the temperature of hospitals and other places in the interests of health. Huntington was a firm believer in the refreshing effects of weekend visits to the sea or mountains and in the value of holidays. He thought that there was an annual wave of energy, and suggested that factories should be run accordingly, with the machinery running slowly in winter, faster in the spring and in May perhaps 10–15 per cent faster than in January. In the full summer it would run more slowly than in May, but not as slowly as in winter: and during the autumn it would run at a greater speed than at any other time of the year. The operatives, it was suggested, would scarcely notice any difference, and would preserve their health and do more work than at present.[3]

Huntington goes further

Climate was regarded by Huntington, in common with many other writers, as a dynamic rather than a static force. He quoted with approval from H. L. Moore's work on economic cycles in which the prices of crops were related to the general trends of prices and

[1] *World Power and Evolution*, New Haven 1920, 47–104.
[2] As noted on p. 102.
[3] *Civilization and Climate*, op. cit., 290.

the explanation of the recurrent trade cycles was sought in the oscillations of weather. Moore formulated a law:

The weather conditions represented by the rainfall in the central part of the United States, and probably in other continental areas, pass through cycles of approximately thirty-three years and eight years in duration, causing like cycles in the yield per acre of the crops; these cycles of crops constitute the natural, material current which drags on its surface the lagging, rhythmically changing values and prices with which the economist is more immediately concerned.[1]

Huntington goes further in saying that statistics collected from 1870–1914 showed that a high death rate precedes (not follows) hard times, while a low death rate precedes a time of prosperity. In short, health is a cause rather than an effect of economic circumstances, and fluctuations in health affect mental efficiency, drunkenness, bank deposits, prices and immigration.[2] Why do people drink? Huntington, whose attitude shows some signs of his background, says they do so in America (social conditions differ in Europe) because they need bracing up, lack the strength of mind to refuse invitations or to resist their own desires: the steady drinker consumes most when out of work (or would if he could afford it). Years of poor health, it was observed, were followed by years marked by heavier drinking than usual, such as 1873, 1888, 1896 and 1911. There appears to be a series of cycles in business affairs, of approximately nine years, varied by shorter cycles each of some forty-one months. Could this be due to sunspots? Having accumulated a good deal of evidence, Huntington is uncertain of the causes of such cycles though he suggests that they may be due to the sun's electrical effect on the earth's atmosphere.

For a long time, the trade cycle was almost an economic cliché. Huntington was sure that ordinary weather cycles were the chief stimuli to invention, thrift and foresight: this does not, however, preclude the possibility that cycles of varied origin lasting several years exist, perhaps related to the quantity of ozone in the atmosphere, and to the varying electrical qualities of the air. The ozone cycle may be reflected in varying mental activity, for which Hunting-

[1] Moore, H. L., *Economic Cycles: their Law and Cause*, New York 1914, esp. 149. See also Huntington, E., 'Climatic variations and economic cycles', *Geogrl Rev.*, 1, 1916, 192–202, in which the work of O. Petterson is discussed with that of Moore.

[2] *World Power*, op. cit., 29.

ton used as an index the circulation of library books. This may cut
across his other theory that low outdoor temperatures were con-
ducive to mental activity. In any case there was conflicting evidence,
for in Calgary, the maximum reading of serious books occurs in
mid-February when the temperature averages 13° F., at Tampan,
New Orleans and Houston in March (62°–66°) and at Panama in
April with an average temperature for the month of 78° and 85° at
noon: therefore the hypothesis of a mental optimum of 40°–50° F.,
to be correlated with book reading, is not entirely satisfactory.[1]
Of all the climatic factors, none seemed to Huntington of greater
importance than the incidence of cyclonic storms, for plants,
animals and man all respond immediately and in diverse ways to
the changes that accompany such storms: the stimulus lay in air
changes due to pressure rise or fall, wind movements, humidity and
rain, temperature, the amount of ultra-violet light reaching the
earth's surface, ionization, the variation in ozone and also of atmospheric electricity. A long section on cycles includes the statement
that

The centre of civilization during the past few thousand years has
shifted not only from warmer to cooler climates, but from those with
few climatic storms and presumably with little of the stimulating
activity of atmospheric ozone and electricity to those with a maximum
of these stimulating conditions. This appears to be one of the vital
elements of the history of civilization.[2]

Summarizing the effects of stimulating storms, Huntington says
that the stormy part of western Europe is the most favoured of all
areas for mental efficiency in our stage of civilization, within a
rough rectangle bounded by Liverpool, Copenhagen, Berlin and
Paris.[3]

Inevitably, Huntington was confronted with the view that race
was a determinant of action, and in 1919 he published a book on
the indigenous inhabitants of America,[4] in which he assessed the
American Indian as possessing a mental capacity between that of
the Negro and the white man. The Aztecs of Mexico had a high
degree of social organization and elaborate religious ceremonies,
but gave little resistance to European conquest: possibly they had

[1] *Mainsprings of Civilization*, New York and London 1945, 354–5, 527.
[2] Ibid., 528.   [3] Ibid., 384.
[4] *The Red Man's Continent*, New Haven 1919.

I

advanced so far under the stimulus of a climate that was subject to storms though dry and warm with summer rainfall. The Iroquois of New York had a vigorous civilization with a strong political organization and lived in a climate where cyclonic storms brought rain at all seasons: the fishermen of the British Columbia coast, even more advanced commercially but comparatively gentle in character, lived in a mild oceanic climate. There were, in short, considerable differences in the habits and mode of life of the various tribes, but those arose from the topography, the climate, the plants and the animals which formed the geographical setting of their homes.[1] Huntington was clearly anxious that all the features of the environment should be considered and his study of the American Indians concludes with a criticism of two respected works, both published in 1903, E. C. Semple's *American History in its Geographical Condition*, and A. P. Brigham's *Geographic Influence in American History*:

> both of these books interpret geography as if it included little except the form of the land. While they bring out clearly the effect of mountain barriers, indented coasts, and easy routes whether by land or water, they scarcely touch on the more subtle relationships between man on the one hand and the climate, plants and animals which form the dominant features of his physical environment on the other hand.[2]

He mentions that in making this criticism, he had followed the lead of W. M. Davis.

Having an abiding interest in evolution, Huntington reached the conclusion that racial characteristics were liable to change under the stimulus of environmental influence, and in any case they were less important than those due to innate individual differences, some of them due to environment. In this case he apparently interpreted the term 'environment' widely, including the social and economic aspects of life as well as those of immediately obvious physical origin. The key lay in heredity, which could only be understood by studying differences among individuals, family stocks, occupation groups, urban and rural types, castes or social classes.[3] Some groups make more progress than others. This problem was faced in a study of the people of Newfoundland and Iceland, which led to the view that the Icelanders were a superior stock, both temperamentally and in

[1] *The Red Man's Continent*, New Haven 118-22.   [2] Ibid., 173.
[3] *Mainsprings*, op. cit., 39.

ability, to the Newfoundlanders, and that they had maintained and even increased their superiority by selection.

. . . the present great advantage of Iceland over Newfoundland can be due only in part to cultural inheritance. Its main source seems to be that the Icelanders, in spite of scanty resources and great isolation, have had the innate capacity to build a much better structure than the Newfoundlanders, even though the latter had much better raw materials, and the help of the mighty British Empire. What we are here studying, then, is an example of the relative effect of environment versus heredity, for both culture and physical conditions are part of man's environment.[1]

This study of groups led to theories of 'Kiths', such as Jews, Parsis, Armenians and others, who had certain qualities, acquired partly by selection, for example among the Jews where many of the weaker elements had perished and the more perspicacious survived. Some Kiths, such as the Armenians, appeared to be indestructible as the survivors of Turkish massacres were in a few years more prosperous than their persecutors. Other groups, spoken of as a 'Microkith', had acquired a world reputation for some attribute, such as military skill in the Junkers: this did not mean that they were inherently warlike but rather that in the circumstances of their day and age they had used their natural efficiency as soldiers. Frequently, however, Huntington points out that every group has people of many grades: not all Icelanders are efficient nor all Newfoundlanders inefficient. Nowhere was selection more clearly shown than among nomadic peoples:

The necessities of the pastoral life demand certain distinct characteristics, among which leadership, bravery, self-reliance, reliability, and adaptability, are vital. If young men who are weak in these qualities remain among the nomads, they fail to get wives or are given girls whom others do not want, and who are often physically deficient so that they have few children. The result is that their kind tends to die out. On the other hand, the young men in whom the requisite qualities are strongly present get the most desirable wives, and are more likely than others to have several. Thus their stock increases. If temperamental qualities are hereditary, as most assuredly they are, it seems almost certain that prolonged biological selection must have tended to give nomads unusual ability as soldiers, organizers and rulers.[2]

Like many of the human geographers of his time, Huntington was conscious of the relevance of migration in history: it could

---

[1] Ibid., 140.   [2] Ibid., 179.

hardly have been otherwise as his first studies in Asia on climatic pulsations had a clear relation to migration movements. Summarizing his work, he suggested five 'laws' of migration;[1] first, migration is systematically accompanied by selection. If forced by conquest, flood or some other disaster, it may at first seem to be non-selective, but the deaths will be greatest among the weaker types. Second, the harder the migration, the greater the selection. Third, as people migrate, so differences in racial, national or social characteristics will lose their sharpness, especially if the movement is long and difficult. All are reduced to the same level on the pioneer fringe, and the Chinese who pushed forward to the limits in Manchuria during the 1920s were very much like the British pioneers in Canada or the German settlers in southern Brazil. Fourth, a wide range of people emigrate, of which examples given include the upper class group who fled from Russia and the criminals shipped from Britain to Australia. Finally, migration is impelled both by conditions at home and those abroad. Huntington considers various groups in different parts of the world, and reaches the conclusion that many civilizations were largely made by immigrants, such as those of Ceylon, Cambodia and Java who 'merely took the culture evolved in north India and added a few relatively minor items'.[2] And some areas owe virtually everything to the quality of their inhabitants drawn from other areas of more favourable climate, such as the British in Queensland and the Americans in Hawaii. But can enterprise survive in such climates? Huntington appears doubtful, for he says that: 'leisurely rest and social amenities get more time than in more bracing climates, whereas such matters as serious reading, inventions, new projects, and the promotion of education, health, and good government get less'.[3]

Some of the theories of life put forward by Huntington will seem absurd to many readers. Worldwide statistics, he claimed, showed that the human species has an annual cycle of reproduction like that of animals, with a maximum of births in the early spring. But in certain groups, such as the upper classes of Britain and Germany and the wealthier classes of the United States, there is a summer maximum of births due to conceptions in the autumn when, Huntington avers, there is a feeling of exhilaration almost everywhere. But the less fortunate classes, including most criminals, insane per-

[1] *Mainsprings*, op. cit., 96–7.   [2] Ibid., 397.   [3] Ibid., 391.

sons and sufferers from tuberculosis, both in the United States and in western Europe, have children born in the spring. There appears to be a right time to be born, for a large number of the great leaders of the world were born in the late winter. On the other hand, at least until the nineteenth century, some kind of physiological endowment still gave Americans and Italians the greatest vigour if conceived in June and born in March, that is, in conformity with the rhythm of man's progenitors. But such an outburst of fecund enterprise in June had its darker side for, having studied the seasonal outbreaks of insanity in Italy, England, the United States and Germany, of suicide in Italy, France and England, and of sexual offences in France and Germany, Huntington shows that such events were 80 per cent more numerous in June than in December.[1] And the warning is added that such violent disturbances might result in handicapped children. In the summer, health was generally more satisfactory than at other times, yet there were obvious social disadvantages and difficulties in the glorious days of June. Still more was this the case in hot countries, for after a consideration of Indian riots Huntington reaches the conclusion that

In the world as a whole the tendency toward lack of self-control in politics, in sex relations, and in many other respects rises markedly in hot weather and in hot countries. This is not the only reason for the frequency of political revolutions in low latitudes, but it must play a part.[2]

The assessment of the relative degree of civilization of countries of the world was perhaps the crown of Huntington's enterprise. Civilization is based on innate capacity, a physical environment with sufficient natural advantages to maintain a rising standard of living, a good cultural inheritance and vigour to use the opportunities that arise. The countries are rated from 100 downwards, and fall into four groups: the highest score in this 'health and vigour' assessment is given to New Zealand (100), as it has people of intelligence and vigour, a climate good for grass, cattle, sheep and man, and the economic good fortune to sell its products on the other side of the world. Other countries included are Australia (98), the white sections of the United States and South Africa, and Canada, all of which have profited from recent selective migration of comparatively high culture, a small population in proportion to natural

[1] Ibid., 273, 319–21, 365–6, 610.   [2] Ibid., 365.

resources and a favourable climate in the occupied areas. Also in this group are several European countries, Norway, Sweden, Denmark, England and Germany. The next group consists entirely of European countries regarded as less favourably endowed, Eire (the Irish Republic), mentioned elsewhere as too wet, Belgium, France, Scotland, Finland, Austria, Italy and Czechoslovakia. The third group has a lower grading in climatic efficiency, and includes four countries, Hungary, Japan, Poland and Bulgaria, with problems of overpopulation, as well as Greece, where the somewhat healthier physical conditions are ascribed to selection during the immigration from Asia Minor following the 1914–18 war. Russia is also included in this third group. Last of all come the least favoured countries of the world, such as Egypt and India with indices of 52 and 45 respectively, and many more for which no reliable statistics exist. These are relatively poor in natural resources, have unhelpful climates, and have had no recent selective immigration of stronger stocks.[1] The various maps that accompany this statement are of relative grades of civilization in the various countries.

The contribution of Huntington

In this chapter a considerable number of quotations from Huntington's work have been given, and it is hard to resist the temptation to give even more, for his writings abound in purple passages, in vivid and stimulating statements. He undoubtedly played with ideas and, fed in youth on the great concept of the unity of nature that could be traced to the Darwinian concept of evolution, he sought to show the universal geographical influence of environment in all stages of civilization. And he was concerned with the totality of the environment. Trained in zoology, he was a highly competent physical geographer as many pages of his books show clearly, but for him the mere study of geomorphology was not enough. In his work on the changes of climate he fought a battle that has since been won, for much evidence, archaeological, historical and contemporary, has been assimilated to show that such changes occur. This does not, naturally, imply that Huntington's views of consequent changes in civilization have met with equal acceptance. To many it will seem that Huntington made such a desperate attempt to be scientific that he ceased to be scientific at all. In searching for general laws

[1] *Mainsprings*, op. cit., 254–9.

on human behaviour he used what data he could find with undoubted enterprise, obviously impressed by the rising interest of
his time in psychology and in eugenics. The faith of his day in the
transmission of hereditary qualities has been somewhat dimmed
since he wrote and the effects of environment are far more complex
than Huntington's more dominating statements would suggest.
Though commonly dubbed a 'climatic determinist', Huntington
was a man of far wider mind than some of his writing indicates.
For the geographer he is of interest for his attempt to study the
totality of physical environment in its influence on man, with
climate as something experienced day by day, hour by hour, yet
over centuries moulding and re-forming the physical earth and
therefore of permanent significance.

BIBLIOGRAPHICAL REFERENCES

As indicated on p. 101, the obituary in *Ann. Ass. Am. Geogr.*, 38,
1948, 38–50, gives a full bibliography and a short life of Huntington.
See also *Geogrl Rev.*, 38, 1948, 153–5 (by S. Van Valkenburg).

CHAPTER 6

Sten de Geer

A PRACTICAL GEOGRAPHER

STEN DE GEER had only a short life, for he died at the age of
forty-seven in 1933. He came from a distinguished Swedish
family, and his father, Gerard de Geer (1856–43) is widely
known for his work on glaciology, though this lies beyond the scope
of this book. Sten de Geer showed in much of his work an intensely
practical approach both in physical and in human geography,
though as the years went on he became more interested in large
areas such as the United States and even the whole world. But his
atlas of population in Sweden, of 1919, a classic example of careful
mapping and local investigation, probably had a greater impact on
geography than such articles as 'The subtropical belt of old Empires',
published in 1928, or his efforts to define the 'Nordic area' of Europe
of the same period. Past work is 'dated', it is said, but in different
ways. The 1919 atlas shows, within its range of expression, the distribution of population in Sweden in 1917 and is, therefore, a
significant document for the historian of the future. The articles of
his later years are interesting mainly as an indication of the geographical thought of the time, though an exception may justifiably
be made for his long article, of over one hundred pages, on the
American manufacturing belt, published in 1927, in which he used
the European manufacturing belt as an analogy for the one that had
grown up on the other side of the Atlantic. Despite the marked
differences in historical evolution, one main manufacturing belt had
developed in each continent. He was obviously aware that there
were signs of expansion in an America still pulsating with immigrant
vitality and moving into possession of vast natural resources, though
he regarded even Denver as too remote for marked growth to be
expected and commented that the Pacific cities were isolated one
from another.

Born at Stockholm in 1886, Sten de Geer lectured at Stockholm
University from 1911 to 1928, and then went to the University of
Göteborg where he died on 2 June 1933. Most of his writing was

PLATE 3. Sten de Geer

done within the sixteen years from 1912 to 1928 for at Göteborg his main concern was to establish the subject firmly in the University. His travels included a visit to Spitzbergen in 1908, periods in Bohemia and Silesia in 1910, in France and Italy during 1910–11, and in the United States during 1922. In America he was obviously intrigued by the wide-spreading thought of Ellsworth Huntington, both in racial problems and on the climatic influences on people and nations. Another influence on de Geer was the geopolitical thought that originated with Kjellen in Sweden and developed mainly in Germany. R. Kjellen's book *Staten som Lifsform* was published in 1916 and appeared in a German translation as *Der Staat als Lebensform* in the following year. Especially interesting is de Geer's work on the 'core' or nuclear areas of states in the 1928 paper on subtropical empires, in which he notes that R. Kjellen's *Stromakterna* (Great Powers) 1905–13 was of value: he had drawn historical data from *Weltgeschichte* by H. F. Helmolt, 1907 (revised by Armand Tille, 1913), and also from various historical atlases, including those of Kiepert, Spruner-Meuke, G. Droysen, F. W. Putzger, Charles Joppen and Vidal de la Blache.

In 1923, de Geer published a paper on the content of geography but this, though interesting and worthy of close study, seems more restricted than the work of the next few years. There are perhaps two stages in academic experience when a worker may attempt to define his subject: one is about the age of twenty-five when a man has sharpened his sword for interviewing committees for various posts, and the other is approximately thirty years later, when there has been time to test the theories confidently held in the dewy freshness of initial academic success. Sten de Geer wrote his apologia at neither time, for at twenty-five he was concerned, very rightly, with the minutiae of research and he did not live to be fifty-five. His paper of 1923 was published when geography was receiving more attention than ever before in American and European universities. And the view of the time was at once local and world-wide. Initially, de Geer's work had been strongly local in tone, but he had begun to write on the new Europe and to attempt, like other workers before him, to devise a world regional scheme. At the beginning of his article, he seems aware that geography must, or at least should, 'serve the present age' to quote John Wesley, for 'now one side of Geography and now another, has primarily caught the interest of a generation; and the general conceptions of the essential character of the new science have varied accordingly'.[1] The initial definition given is that geography is 'the science of the present-day distribution pattern on the surface of the earth',[2] with the understanding that the past need only be considered in order to understand the present. It could here be argued that all study of past time illumines the present, however remote it may at first appear, but de Geer thought that a stress on 'the present day' implied a necessary limitation, and was not apparently in sympathy with the sweep through historical time then characteristic of the work of French and other geographers. Later, however, he used historical data from many periods in his study of subtropical empires.

The essential purpose of distribution mapping was to build up a synthetic regional view of the world. De Geer made a distinction between absolute and relative distributions. The absolute distribution is the appearance or non-appearance of objects within a region: the relative distribution is expressed by height or depth, density, value, force or time. Further, objects exist in relation to one another, and by the comparative study of distributions it was possible to recognize geographical regions and provinces. Geography is concerned essentially with distributions in space, and in method is abstract and non-material, for its purpose is to study general qualities or circumstances in material objects. Nor is it unique in this, but rather a general science like statistics, mathematics, philosophy or even history, all of which deal with heterogeneous materials. Other Scandinavian geographers were making definitions at this time, and of these de Geer quotes with approval that of J. G. Granö of Finland (in 1912) that geography dealt with 'the earth's surface from the standpoint of distribution and interaction'.[3] The idea not only of the simple, or absolute, distribution occurs here, but also that of the interaction of phenomena, or relative distributions. And this introduces the concept of synthesis, for one distribution can only be understood in relation to another, and the character of the relationship distinguishes one area from another. This led de Geer to say that 'geographical provinces and regions form a synthesis of characteristic complexes of important distribution phenomena within limited parts of the earth's surface'.[4] And with this conclusion he adds to his initial simple definition, given above, of 'the science of

[1] *Geogr. Annlr*, 5, 1923, 1.   [2] Ibid., 2.   [3] Ibid., 7–8.   [4] Ibid., 10.

the present-day distribution phenomena (or pattern) on the surface of the earth', the statement that geography 'aims at a comparative and explanatory description of the characteristic complexes of important distribution phenomena—geographical provinces and regions —which occur on the earth's surface'. He looked to the creation of a world regional scheme, and finished his article with a division of the land area into twenty-seven regions and of the sea into seventeen; like many geographers of his time he paid far more attention to the seas than is now usual. Many other geographers had also evolved world schemes such as those of Hettner in 1908, Banse in 1914 and Seiger in 1921; few geographers are now concerned with world schemes of regionalization and many would regard any attempt to create such a scheme as impossible. But is this because we know too much or is it due to lack of courage?

As a cartographer de Geer showed a degree of common sense that is most uncommon. He said that

the old descriptive geography seldom based its pronouncements on a sure quantitative basis; nor, indeed, does the new genetic geography which extends its view backwards toward the causes and forward toward the effects of the facts treated. It appears to be different, however, with the new descriptive geography; and we may hope, so far as source material permits, that it may develop into a genetic-quantitative geography, charting in a clear way its detailed results with regard to chronological as well as spacial relations.[1]

He recognized that no map can possibly show everything that exists on the ground, but that some give a helpful if partial synthesis, such as American topographical sheets showing relief and other landform features in brown, water in blue and human distribution in black. Wide as the range of information may be, it is never the entire picture. He experimented with methods of representation, such as shades of the same colour, line shadings, dots or a network of dots or points and even pyramids and cubes; perhaps he is most widely known for the globe symbol used for towns in his population atlas of Sweden. He was the first to use the dot method for the detailed population maps of Sweden, combined with a brown layer colouring which brought out the relative intensity of settlement in rural areas. Here a distinction is made between the absolute and relative methods, as in his other work. The absolute method related directly to the

[1] *Geogrl Rev.*, 12, 1922, 72.

observed phenomenon; for example, a dot to 100 persons, a globe
varying in size for the town populations.[1] But the relative method
refers to area: for example, a certain area is sparsely populated,
another densely settled, so tints of different intensity cover each area
accordingly. Both are generalizations, for the globe does not cover
the exact area of the town neither does the light brown wash for
sparsely peopled areas effectively exclude some areas that are uninhabited nor does it accentuate others quite thickly settled, such as a
few acres of houses in a village. Yet for a general picture, the dot
distribution maps of population proved to be a major advance in
cartography, copied with modifications by many later workers including the present author. C. B. Fawcett, in an obituary of 1933,[2]
noted that de Geer had in the same year presented to a committee
of the International Geographical Union a proposal to produce a
world map of population on the 1 : 1,000,000 scale, using a dot
method. So far this has not been achieved, though it was discussed
again, and appropriately, at the International Geographical Congress at Stockholm in 1960. At least the seed was sown in 1933,
even if the harvest is delayed.

In the remainder of this chapter, attention is given first to the
detailed earlier work of Sten de Geer and then to the more general
work that followed later in his life. Like many of his contemporaries,
de Geer had a breadth of mind reflected in a penchant for generalization that many of his modern successors would not emulate. Some
would view his work on the Nordic area as an interesting period
piece, derived partly from the over-confident recognition of 'racial
types' characteristic of the 1920s. And though the recognition of
regions has not ceased to be a reasonable geographical enterprise,
by no means all workers would regard it as practicable. Such a view
is expressed by W. R. Mead in writing of the *Bergslagen*, admittedly
one of the most difficult areas of Sweden to define:

An early and almost classical statement on *Bergslagen* is that of Helge
Nelson. Nelson, like Sten de Geer, has claimed that the region is the
summit of geographical consideration. The study, written in the field
by its young author (in 1913), attempted to recognise the essentials of
Bergslagen in a limited area, and it married historical appreciation
with field observation. Yet its impact upon the geographer of forty
years later . . . is much the same as the impact of Mendelssohn upon
the ears of a contemporary musicologist. . . . The contemporary

----

[1] Pp. 129–34.   [2] See p. 154.

geographer . . . is aware that the varying interpretation of the economic region . . . represents a groping towards rather than a grasping of its essentials. Secondly, he is aware of the formidable task of satisfactorily undertaking a survey with . . . current statistics, cartographic devices and observational teamwork.[1]

The local studies

When hardly twenty years of age, de Geer published studies of two rivers, the Dal (near the Alfkarleby falls) and the Klar. A map shows the past courses of the Dal, the areas subject to erosion, strandlines beside the river and the effect of ice in the first days of January 1906. The paper on the Klar is rather fuller, and includes data on the depth of the river, an analysis of its breaks of slope, terraces and meander forms, and a study of the area where the Klar joins lake Vanarn showing deltaic areas of various dates. Some years later, the interests already apparent were extended to lakes and harbours; in 1912 de Geer gave an analysis, with sounding lines, of three lakes west of Orebro (on the Närke river) and in the same year he wrote on the landscape near the Dal river, noting the existence of meadow, forest or brushwood; this article includes a map on the 1 : 6,000 scale of a section of the Dal, showing erosion, depths of the river at various points and areas reclaimed. All this work was of value in a country subjected to glaciation as strongly as Sweden, and some modern Swedish workers are now exploring the river depths as frogmen. In 1913, a short paper on a Spitzbergen glacier appeared with some geological treatment and photographs showing the calving off of icebergs, apparently one fruit of the visit to Spitzbergen. Much of this early work provided a basis for the regionalization of Sweden, which came later on both a physical and a human foundation.[2]

Meanwhile, in 1908, de Geer published an article on Gotland with a population map that was the precursor of the famous population atlas of 1919. On a map of 1 : 300,000, he showed the population distribution with one dot to ten people, with the limestone uplands distinguished as a local feature, and on an inset map on the 1 : 900,000 scale the populated areas were shown. The 1919 atlas uses the census of 1917, and includes twelve maps, each on a scale of 1 : 500,000, with one dot to every 100 people, groups of dots for places having

---

[1] Mead, W. R., *An Economic Geography of the Scandinavian States and Finland*, London 1958, 277–8.

[2] References are given to these papers at the end of this chapter.

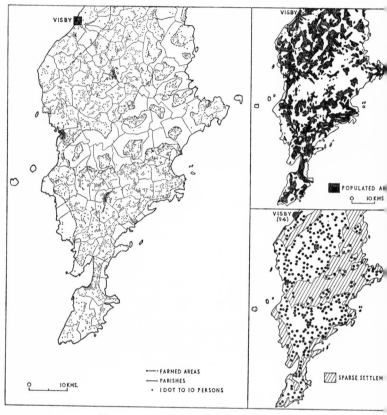

Fig. 12. Population maps of Gotland

200–5,000 people, and the globes to which reference has already been made[1] for larger places; in villages the dots are placed haphazardly to indicate the loose clusters, but in small towns they are arranged in symmetrical rows to show a more continuous and orderly form of settlement. There is a graded series of brown tints of deepening intensity for areas of sporadic settlement, thinly peopled and densely peopled respectively; the areas entirely uninhabited are left blank and a red shading, with spot heights, is used for areas over 400 m. high.

[1] See p. 127.

The larger map shows Sten de Geer's use of one dot to ten persons mapped by each parish: the upper map was given as an inset, and showed the inhabited areas of the island. For comparison, the lower inset map illustrates the further generalization of the 1919 atlas, with 1 dot to 100 persons.

The larger and the upper inset maps are re-drawn from a map in *Ymer*, 28, 1908, opp. p. 252. Note that de Geer was not the first to use such methods of mapping: see A. O. Kihlman, 'Om naturliga områdero användning i statistiken', *Fennia*, Helsingfors, no. 1, 1897–9, 46–59, in Swedish with a French summary. Kihlman's illustrations (opp. p. 84 in op. cit.) shows that he used one dot to ten persons and then calculated the density for each territorial division. In the *Atlas de Finland*, Helsingfors 1910 (the second edition of the national atlas) there is a map (plate 26) for various selected areas of Finland (1905 figures) with one dot to ten persons.

De Geer was acutely aware of the dangers of generalization. In his article 'On the definition, method and classification of Geography', *Geogr. Annlr*, 5, 1923, 1–37, he includes diagrams similar to those opposite, showing the distribution by dots of population and the populated area by shading. He adds two more illustrations showing how generalization of this populated area may give far larger areas than that actually inhabited (p. 17), and adds with approval (p. 18) Ratzel's comment on 'the importance of investigating the gaps'. If this seems too obvious to mention, a critical study of the maps in almost any geographical writing or in any atlas will show a devastating degree of generalization, some of which may be defensible or even necessary, but not all. The illustration opposite shows that the 1919 atlas generalized the detailed work of the 1908 paper and the reader can assess for himself how much has been lost (or gained) in the process. The real point is that de Geer had in mind the actual landscape: he evoked the Swedish rural scene with its interpenetration of farmlands and forests or, in the case of Gotland, some limestone pastures. Visby had 9,400 people.

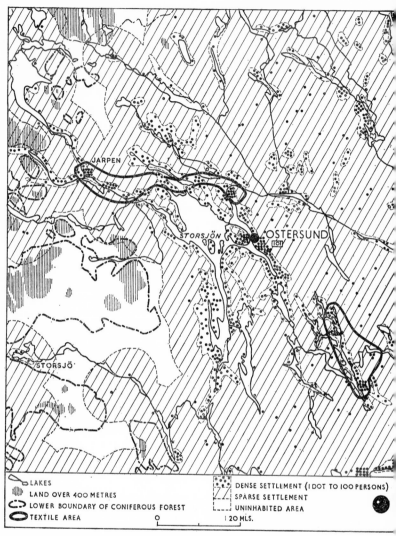

LEGEND:

LAKES

LAND OVER 400 METRES

LOWER BOUNDARY OF CONIFEROUS FOREST

TEXTILE AREA

DENSE SETTLEMENT (1 DOT TO 100 PERSONS)

SPARSE SETTLEMENT

UNINHABITED AREA

0   20 MLS.

Fig. 13. Population distribution in part of Jamtland

Redrawn from the atlas discussed in the text, this map shows the range of information given in the atlas. It was not possible to show a globe for Ostersund (10,100 people), but one has been inserted in a spare corner at the foot of the map. For many years globes had a decided vogue. The population in hundreds was shown.

In itself the atlas was a contribution to the regional geography of Sweden, and de Geer includes the upper limit of cultivation mapped in 1872, the limit of deciduous forests of 1903, and the upper limits of private and state forests. Other information of a regional character is given by the mapping of sandy glaciated areas with marked ridges separately shown, areas of settlement on plateaus and on hilly areas of a plateau character. Rivers having basins of more than 1,000 square kilometres are separately distinguished from the others and major water partings are indicated, with the upper limit of marine deposits from former seas. A further key to local regionalization is given by the boundary of the Silurian limestone areas; on a larger scale limits are given for the uplands, for Norrland and for South Sweden. Areas notable for the timber industry, textiles, glass and the very limited coalfields are included, and there is information on roads, railways (actual and planned), canals, steamship routes and fishing grounds. A reviewer of 1920, though praising the work, notes that one dot to 100 people, which would mean twenty or more households, is hardly satisfactory in a country having large areas sparsely populated. Here compromise is inevitable, for the symbol should remain constant over the entire country, and one dot to 100 people gives a tight network of dots in the more closely settled agricultural areas, particularly those with some admixture of industrial and suburban settlement.

The atlas is fascinating to study and must have been laborious to compile; it is accompanied by a volume of text which is a substantial contribution to the regional geography of Sweden. In 1922 de Geer discussed his work,[1] and shows that, if the towns and industrial areas were excluded, the rural density was generally 50–100 per square kilometre, except in a few areas such as Upper Dalecarlia, in North Sweden, where it was 200 to the square kilometre. There the old villages loved by the people survived, and production was maintained at a rewarding level by the intensification of farming, forestry and rural industry. But in most areas the total population had been declining from 1865, when it reached its maximum, partly through emigration, as in Varmland, and partly through the introduction of labour-saving machinery. De Geer regarded the population mapping as part of a wider scheme and spoke of the need to relate this

---

[1] *Geogrl Rev.*, 12, 1922, 72–83.

K

distribution to natural and human geographical factors. On its application to human activities, he commented:

The population map . . . has a practical value . . . in questions of readjustment of boundaries of administrative divisions or of social organizations, of the establishing of public or private institutions, of lines of communication, of the stationing of officials, of the selling of goods, of educational propaganda, of organization of traffic in peace and mobilization in times of war—in short, in any matter where it is necessary to know the number and grouping of people.

The first signs of an interest in urban geography were shown in an article of 1912,[1] in which de Geer deals with various major ports on the Baltic and includes a map on the 1 : 1,000,000 scale showing the central area ('centrum') of Stockholm, Helsinki, Leningrad (St Petersburg), Riga, Reval, Königsberg, Danzig, Stettin, Lübeck and Kiel. There is also a dot map of the population of Stockholm ascribed to another worker (N. Victurin) and the references include a paper by the American, Mark Jefferson, on the 'anthropogeography of some great cities'.[2] Fifteen years later, in 1927, a second paper appeared on the Baltic harbours, in which a comparative study was made of their use in 1912–13 and 1923–4.[3] This showed an increasing measure of concentration on the larger ports. Dot symbols were used to show the relative volume of trade at each port and the intensity of the rail services was also indicated. Plans of the ports themselves show that there were four main types—quays with piers, quays with deep water such as Helsinki, breakwaters sheltering basins such as Reval, and ports in rivers and estuaries.

From these beginnings, de Geer turned to a consideration of the geographical aspects of Greater Stockholm, on which he wrote an article published in 1923.[4] The main natural advantage of the site is the crossing of the north–south land route and the east–west

[1] *Ymer*, 32, 1912, 41–87.
[2] Jefferson, M., 'The anthropogeography of some great cities', *Bull. Am. Geogr. Soc.*, 41, 1909, 537–66.
[3] *Svenska Hamnförbundet Österjöhamnar nas Geografi*, Stockholm 1927.
[4] *Geogrl Rev.*, 13, 1923, 497–506. This article is a summary of a study of Stockholm in Swedish (ref. p. 155) dated 1922.

The intricacy of de Geer's observation is well shown by this map, which includes some of the information given in the maps copied from the Swedish originals into *Geogrl Rev.*, 13, 1923, 502.

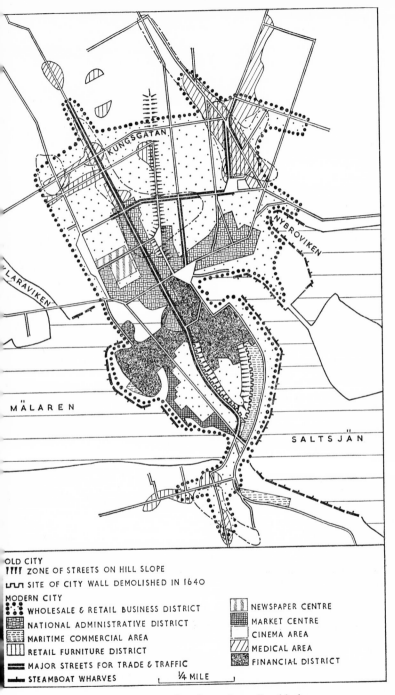

OLD CITY
|||| ZONE OF STREETS ON HILL SLOPE
⊔⊔⊓ SITE OF CITY WALL DEMOLISHED IN 1640
MODERN CITY
⦙⦙⦙ WHOLESALE & RETAIL BUSINESS DISTRICT
▦▦ NATIONAL ADMINISTRATIVE DISTRICT
▦▦ MARITIME COMMERCIAL AREA
|||| RETAIL FURNITURE DISTRICT
▬▬ MAJOR STREETS FOR TRADE & TRAFFIC
▬◄ STEAMBOAT WHARVES   ¼ MILE

▤ NEWSPAPER CENTRE
▨ MARKET CENTRE
|_| CINEMA AREA
▨ MEDICAL AREA
▨ FINANCIAL DISTRICT

FIG. 14. Sten de Geer's work on Stockholm

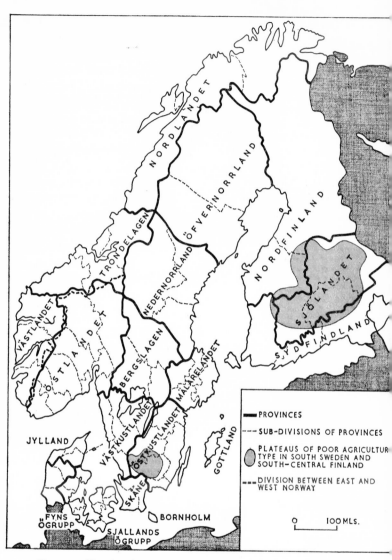

PROVINCES

SUB-DIVISIONS OF PROVINCES

PLATEAUS OF POOR AGRICULTUR
TYPE IN SOUTH SWEDEN AND
SOUTH-CENTRAL FINLAND

DIVISION BETWEEN EAST AND
WEST NORWAY

0   100 MLS.

FIG. 15. De Geer's suggested political divisions in Scandinavia

De Geer proposed to reduce the number of provinces in Sweden from
twenty-five to eight, designed as suitable units for economic life and

water route from the Baltic to the extensive fresh-water Mälaren
lake. The old city, now the main government quarter, grew on an
island crossed by an esker, but the town now has the main commercial quarter on the north side, that is on the mainland. There were
two great periods of modernization, in the 1630s when Sweden was
ruled by a powerful king and was a major European power, and
from the 1860s. Sten de Geer uses historical material in this paper
with a clear eye for the illumination of the present, and calls attention to traces of the past in the existing town. He is clearly aware of
the dynamic and changing character of an obviously prosperous and
growing town. He analyses the qualities of the central business district, which lies outside but adjacent to the main governmental
quarter. And he shows the immense variety of the central business
district which includes the main commercial core, within which he
divided the streets into three groups measured by the amount of

---

transport. Each was to have much the same population, except in
Norrland; the figures for 1915 were:

Öfvernorrland	340,000
Nedernorrland	645,000
Bergslagen	715,000
Mälarelandet	1,101,000
Östkustlandet (Östgötlandet)	1,054,000
Västkustlandet (Västgötlandet)	1,088,000
Skåne	711,000
Gottland	55,000
	5,713,000

In Sweden, the five southern provinces are based on geomorphological characteristics: the southern boundary of Norrland is the
northern limit of the oak, fruit trees and pure agriculture; it is also
the southern limit of heavy snow, Arctic winters and lumbering. In
Norway there is a clear division between east and west. Whenever
possible watersheds are used as boundary lines. In Denmark, de Geer
regarded his scheme as necessarily tentative, but it was based largely
on soil types.

Originally published in *Ymer*, 38, 1918, map (opposite) on p. 44,
table on p. 27. The original article includes several maps contributory
to the main regional scheme: the final composite map of regions was
reproduced in *Geogrl Rev.*, 11, 1921, 143.

pedestrian traffic and the percentage of the street front given to window space for retail shopping. In this area, there are retail and wholesale markets, and an interesting quarter having a large number of furniture shops beyond the main retail centre as there space is too expensive for large showrooms. Facing the Baltic, isolated from the central business districts, there is a quarter of shipping offices and chandlers. Distinct newspaper, cinema and medical quarters are also recognized. Attention is given to transport, and particularly to the tramway stations on the streets. Though a city built on a topographically varied site, Stockholm has developed four concentric zones; first, the banking centre; second the remaining city district; third, the densely built residential zone, and fourth the sparsely built suburban belt. Elaborate techniques of urban geography have been developed since 1923, and some fine work has been done on Stockholm by W. William-Olsson,[1] but de Geer's paper of 1923 is perspicacious and informative and he was obviously in touch with the enterprising American urban geographers of the time.

Sweden perhaps offers fewer problems of regionalization than many parts of Europe and de Geer's work on population was itself a suitable preparation for an assessment of its regional geography. At present, regional studies based on some central theme, such as population distribution, are favoured and regarded as 'new' or 'modern': de Geer used this method some fifty years ago. An initial paper of 1918[2] suggested a new definition of the provinces of Sweden on geographical lines, taking note of the physical features, naturally including the lakes and rivers, the biogeography, historical development and population distribution. Each province would be a unit for transport and commerce, and each approximately equal in population. Two provinces would cover the northern third of the country, two the central area, including a large Bergslagen and a smaller Mälarelandet based on Stockholm, and three the southern part of the country of which Skåne was to be the smallest. Gotland was to be a separate island province. Each province would be subdivided, and de Geer maintained that the provinces would be suitable units for judicial, military and ecclesiastical control, and well suited for the organization of postal services, the railways and general commerce. A discussion of a similar rearrangement of Denmark, Norway and Sweden is given. Proposals to make modern,

[1] See p. 152.   [2] *Ymer*, 38, 1918, 24-48.

logical administrative divisions were made in various countries at this time, and many geographers were interested in such matters. In Sweden, Helge Nelson at once put forward an alternative scheme[1] based on fundamental geographical principles; he apparently regarded de Geer's work as based on a compromise.

In 1926, de Geer produced a regional scheme for North Sweden[2] based on landforms, which distinguishes the highlands of the north from the two premontane areas, which are described as a massif in the north and a plateau from Jamtland to Herjedal. To the east again, there are premontane plains which include the Silurian plain of Jamtland around the Stoers lake, and the similar plain around Lake Siljan. In Västerbotten, there are wide level plateaus, varied by occasional quartzite hills and intersected by valleys with fiordlike lakes. The coasts are varied, and include large areas of finely dissected Archaen plains, a few cuestas now dissected and small areas that are smooth plains, notably around Umea. The main scheme is further subdivided and so too is the scheme of 1925[3] made on a basis of population distribution. This divides Sweden into seven main units: the coastal lowlands of the south with Öland and Gotland, areas of more continuous settlement than is general in Sweden; second, Småland, an area less generally populated; third, the belt of settled country from Stockholm to Lake Vänarn, and along the line of the Gota, largely lake deposits and including areas of considerable agricultural value; fourth, the Bergslagen, which de Geer defines on a relatively lavish scale; fifth, the valleys in the mountain area extending from the west of Lake Vänarn northwards to the Finnish frontier; sixth, the morainic area from Lake Siljan through Jamtland and northwards generally parallel to the regions noted above; and finally the mountain areas, virtually unpopulated.

In fact, the relationship of the settlement pattern in Sweden to the physical features of Sweden is at least discernible, and de Geer was well qualified to show this connection between physical and human features. But he never presented an 'all-purpose' regional treatment for Sweden, and it is idle to ask whether he would have done so had he lived longer. Other schemes of regionalization have

[1] *Geogrl Rev.*, 11, 1921, 143–5, for a discussion of the schemes of de Geer and Nelson.

[2] *Geogr. Annlr*, 8, 1926, 125–36.

[3] *Ymer*, 45, 1925, 392–415: later published in German, see refs. p. 155.

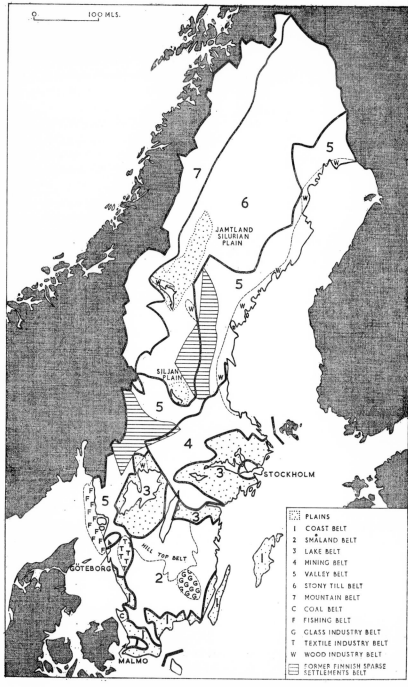

FIG. 16. A regionalization of Sweden

been given by later writers, such as that by M. Zimmermann in the *Géographie Universelle* and the scheme of morphological regions by K. E. Bergsten.[1] Though not identical with the scheme of de Geer, they bear at least some resemblance to it: separate regional systems are used for agriculture and industry, and this perhaps emphasizes that no all-purpose region can be provided with conviction. Even so it may be that Mead (v.s.), in drawing attention to the varied definitions of the Bergslagen given by various writers, and even by de Geer at different times, has drawn attention to an area offering special problems of definition.

This chapter is not a biographical study of de Geer, but the reader may have noted his propensity to write on a theme and return to it after several years. He cast his bread on the waters and it returned after many days. After the 1914–18 war, he became increasingly interested in the wider international aspects of geography, like others of his time. He was also influenced by some of the American geographers, many of whom were raising this subject on a broad front. But it may seem that his most enduring work was that on Sweden and that on the world beyond his homeland ephemeral though indicative of an interesting stage in the wider development of geography.

The wider world

The revision of the map of Europe by the Treaty of Versailles in 1919 induced de Geer to publish a fifty-page article on the subject in *Ymer*, 1920, which was expanded into a book, *Det Nya Europa*, 1922. The paper is a general survey of post-war Europe, noting the distribution of population, the type of government of each state (monarchical, republican, soviet) the prevailing religion, and especially the language or languages used; in these works, de Geer's concern was to present a picture of the new Europe to Scandinavian readers. Two works of the later 1920s show an interest in the major regionalization of Europe on lines of broad general principle, based

---

[1] Somme, A. (ed.), *A Geography of Norden*, Oslo, 1960, 295–99: Bergsten makes acknowledgments, including the Atlas of Sweden.

Fig. 16 shows another scheme suggested by de Geer for the entire country: it is discussed in the text. Redrawn from the original published in *Ymer*, 45, 1925, 403, where it is dated 1920.

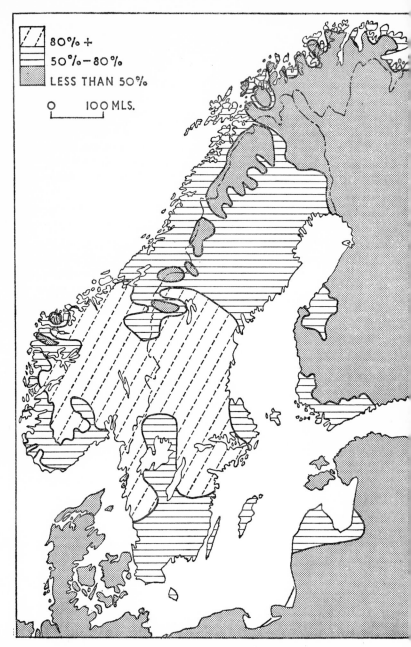

FIG. 17. Head form characteristics in northwest Europe

This map shows the percentage of long-heads, that is, having a cephalic index below 80. It appeared in the 1926 atlas, *The Racial Characteristics of the Swedish Nation.*

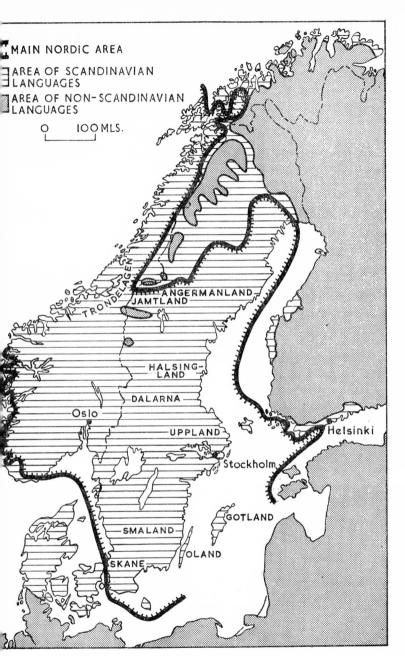

FIG. 18. The main 'Nordic area'

Like fig. 17, this is based on the 1926 atlas. The 'Nordic area' is
defined by criteria discussed in the text, and the Scandinavian
linguistic area is also shown. The material presented in the 1926
atlas also appears in 'Das geologische Fennoskandia und das geographische Baltoskandia', *Geogr. Annlr*, 10, 1928, 119–39 (in German).

on physical geography and on racial distributions. The paper of 1926[1] on 'the kernel area of the Nordic race within northern Europe' is described as 'an attempt at synthetic mapping on the basis of three racial characteristics', mean stature, eye colour and head form. De Geer notes that

the map as inspection and study material is an exceedingly useful form of scientific synthesis. . . . Ethnographical maps demonstrating the distribution of various phenomena have long been in general use. Attempts to show on a map the distribution of race characters such as stature, head form or eye colour, are newer.

It is also stated that the maps were made not only for their geographical interest but also to encourage further work on such problems by anthropologists.

Data on conscripts in the Scandinavian countries with Iceland was used to show that three height zones appeared to exist: first a Russo-Lappic zone of low stature, 156–164 cm. (61–65 in.), second a central, or medium stature zone, 164–168 cm. (65–66 in.), which was designated as 'Karelio-Germano-Baltic', and third an inner high stature, the Scandinavian, zone, 168–176 cm. (66–70 in.), which covered much of South Norway including the Oslo fiord area, but not Jaederen or a fringe on the west coast. In Sweden, this high-stature zone spread from the Oslo fiord area to Göteborg, across Småland, but not Skåne, to the east coast of Stockholm, and then northwards in a belt to Lulea with an inland expansion to Jamtland. A high degree of uniformity was noticed in eye colour, as blue, grey or mixed shades were found in 95–100% in the men over much of Sweden and also in Latvia and Lithuania. In Denmark, however, this range of eye colour was less prevalent, averaging 88·9–92·6%, and on this and other material de Geer postulated that one could recognize three distinct areas, Scandinavian central, with 95%+ of the eyes blue, grey or mixed, transitional areas of 90–95%,

[1] This appears as a chapter of *The Racial Characteristics of the Swedish Nation*, published by the Swedish State Institute for Race Biology, as *Anthropologia Suecia*, 1926. De Geer's paper is on pp. 162–71: he also wrote in the same work, 'A short survey of the geography of Sverige' pp. 53–5. The Institute had been founded in 1925, but the first issue of *Anthropologia Suecia* appeared in 1902. On p. 4 of the 1926 volume it is said that 'Skandinavien (Sverige, Norge, Danmark, Finland and Island) comprises a cultural and for the greater part also a linguistic entity. . . . Anthropologically we are not—nor, for that matter, are any of our countries—a unity; everywhere may be found more or less race-mixture.'

and other areas, including Slavic with 80–90%, with Lappic 65–80%. Head form was based on the cephalic index, and the long-headed types with c.i. below 80 were regarded as 'Nordic'. De Geer mapped the proportion having the 'Nordic' characteristic, and made a series of six zones by percentages, best expressed in tabular form.

Zones of the Nordic Area

Per cent with c.i. below 80	Area	Name of Zone
80–100	Middle Sweden, South Norway	Scandinavian inner
70–80	South Sweden, North Norway	Scandinavian intermediate
50–70	Swedish (speaking) Finland, Southwest Norway, Latvia	Scandinavian outer
30–50	Remainder of Finland, Jaederen of Norway, areas round Baltic, including Denmark	Scandinavian transitional

The areas beyond these, with a lower percentage of long-heads, included the Slavic, with 10–30% and the Lappic with 0–10%.

On the basis of these three indicators, which he calls race isarithms, de Geer works out a cartographic synthesis, using carefully considered limits for each; a height of 172 cm. (68 in.), 95% blue, grey or mixed eye colours; and 80% long-heads with c.i. under 80. To bring out the marginal belts, lower limits were also used: height 168 cm. (66 in.) and 170 cm. (67 in.), 90% eye colour and 70% or 75% long-heads. Correlation of this material led de Geer to the conclusion that

The mapping of these race characters shows such a degree of coincidence with the Scandinavian language area, and race isarithms and language limits so great a conformity, that one leans toward the conclusion that the language area in this case is not entirely without significance in mapping out the distribution of the Nordic race.[1]

This is clearly a cautious statement, but de Geer goes further. By using eight lines, three for height, two for eye colour and three for long-heads, he grades the 'Nordic' region and finds that the fullest development of Nordic characteristics, the 'kernel area', consists of Sweden from the plateau of Småland in the south to Angermanland

[1] Ibid., 170.

and Jämtland and thence to North Norway. A similar prevalence of strong Nordic characteristics is found in Iceland and in the interior fiord communities of the Sogne and Hardanger, whose inhabitants, it is noted, to a large extent colonized Iceland. The main kernel area corresponds to the central and broadest part of the Scandinavian peninsula. A 'pronounced Nordic Region' area covers the greater part of the fiords of Norway, with lower Lappland and Våsterbotten in Sweden, the Åland and Åbo (Turku) skerries in Finland, west Latvia, Blekinge and Akavie in Sweden and the Danish-owned island of Bornholm. A less pronounced but still clear Nordic racial area covers North Norway, the west coast of Norway south of latitude $64\frac{1}{2}°$, Denmark, the east of Latvia, and a large area in Finland beyond the area of Swedish speech but having Swedish place names. The outer Nordic area covers most of Finland, Estonia, Lithuania and parts of Jutland in Denmark. De Geer also notes the spread of Scandinavians beyond their homeland, not least to Great Britain.

Interesting as the conclusions are, de Geer shows an awareness of their limitations when he noted the need for 'an early elaboration . . . with the aid of better, far greater and, as regards geographical localisation, better defined observational material'.[1] In the 1928 article, de Geer uses various maps from the *Nordisk Världatlas* published in Stockholm in 1926, giving a definition of Fennoscandia, a term first used by W. Ramsay in 1900.[2] These maps include one showing the primitive physical nucleus of Fennoscandia with the mountains folded over it, another of the North European peninsula, a third covering Fennoscandia with its bordering morainic areas and a fourth which draws a boundary to the east of the White Sea, including Estonia, parts of Latvia and Lithuania, with Copenhagen and Jutland north of the Mariager fiord. There are also maps showing the distribution of the 'Nordic race', obviously based on the 1926 material discussed above, and the areas of Swedish, Norwegian, Icelandic and Danish, but not Finnish, speech. The area of Protestant Christianity is also shown, with the spread of the Danish and Swedish states during 2,000 years: a map showing the expansionist phases of Sweden is also given. In this work, various lines of evidence are pieced together to define Fennoscandia.

[1] *Anthropologia Suecia*, 1926, p. 171.
[2] *Geogr. Annlr*, 10, 1928, 120–39.

American influences, even the work of Ellsworth Huntington on climate, are reflected in a paper of 1928 that carried de Geer in thought far from his northern homeland in dealing with 'the sub-tropical belt of old Empires'.[1] Having seen the many maps showing the expansion of states and empires, he says that there is a need for 'reduction maps' also, and maps showing the duration of political control. What, for example, is China? Some authors, such as Roxby in England, gave the answer that it was a 'civilisation rather than a nation', having control over a varying range of territory. China has held the lowland of North China for more than 3,000 years, and in periods of greatness it was one organized empire surrounded by a girdle of dependencies, many of them on the west in areas traversed by the historic trade routes to India and the Mediterranean. But Burma, Assam, Nepal and Kashmir were Chinese only during the last great period of the empire, and the Amur countries with Outer Mongolia merely for 300 or 400 years; and the Chinese Empire had extended into the temperate lands of East Asia for comparatively short periods only. The abiding nuclear area, the core, is the lowland of North China. Some students of China have found that it has a discernible geographical pattern, far more so than India, which de Geer considers in detail without reaching any clear conclusions. Other past empires with long periods of greatness include Egypt, which, depending on control of the irrigable valley and delta of the Nile, lasted from about 3,200 B.C. to the Persian conquest in 525 B.C.: Memphis was the capital for at least 1,000 years, Thebes for at least as long, and Saïs in the delta during the later phases. Movements of the capital were regarded as of particular interest by de Geer in a comprehensive study that includes the Japanese, Byzantine, Turkish, Carthaginian and Roman empires.

To many modern students, these broad sweeps through historical time and geographical territory will seem so superficial as to be virtually worthless, but some graduates of the early inter-war period will remember the courses that included a survey of past civilizations for whose territorial range one studied the chastely-coloured plates of the Vidal-Lablache Atlas. De Geer makes some interesting generalizations. First, most of the old empires are in a long row across the Old World, from the Pacific to the Atlantic, between 25° N. and 45° N. This was especially true of the capitals and the

[1] *Geogr. Annlr*, 10, 1928, 205–44.

nuclear areas, though some of the empires had peripheral areas beyond these limits: the only capitals of major empires south of 25° N. are Calcutta and Medina. Second, the empires were associated with the great Old World east–west belt of mountains, which gave irrigation water and rivers charged with fertilizing silt such as those of China. Third, every important centre had at least one or two months of cool or at least 'temperate' weather: de Geer followed Huntington in thinking that a change of weather during the year was necessary to maintain the energy of the leading citizens, and no doubt of others. The sub-tropical belt was defined as 'the wide belt between the northern limit of the vine-growing to the north and the isotherm of a fairly cool January (60° F.) to the south. . . . The central line . . . (is) . . . the northern limit of palms.'[1] The relative vigour of peoples within the subtropical belt gave them such power of resistance that modern great powers had more success as colonizers in tropical than in subtropical areas.

Due to the advance in the north of science and technics, with an increasing veneration for the spirit of enterprise, leadership had passed in modern times to states beyond the subtropical belt, though with the inheritance of much of their traditions and civilization. And there was a need to study the various aspects of political geography afresh: de Geer comments that 'the old political geography has been succeeded by a new human geography',[2] with rational means of investigation. The 'old political geography' presumably meant the tiresome catalogue of states with their subdivisions and towns, and de Geer's opinion was that the 'new' human geography could fructify modern political geography by its study of the grouping and distribution of settlements, roads and other results of human activity visible in the landscape. The distribution of population was of basic significance, even though man himself was less visible in the landscape than his works: de Geer went further in his hope that more attention should be given to the biological and cultural properties of the population, its organization for co-operation in war and peace, for common irrigation in some societies, even for its religious values. The 'new' political geography was mainly an investigation of states and other political organizations, as well as of religious and other social features in communities.

[1] *Georgr. Annlr*, 10, 1928, p. 243.
[2] Ibid., 205.

De Geer was, as mentioned earlier,[1] interested in Kjellen's work on the great powers which analysed 'the internal character of certain states and their intensity as organizations'.

World distributions were the theme of two maps prepared by Sten de Geer for the *Nordisk Världatlas*.[2] To the first edition he contributed a map of world population with one dot to 1 million people and four grades of shading for relative density, on a scale of 1 : 70,000,000. The second map deals with the world distribution of religions, Christian, Jewish, Mohammedan, Buddhist, Hindu, Animist; and of these the Christians, Mohammedans and Buddhists (including Confucianists) are split into sections. The language distribution is included and divided into major groups, such as Norden, which covers Scandinavia, Finland and the British Isles; Romance, including Portugal, Spain, Italy, France and Belgium; and Germanic, Central Europe, with Holland, Luxemburg, Germany, Switzerland and Czecho-Slovakia. He estimated that English was spoken by more people (152 million) than any other language except Hindi (240 million). The maps and an expanded text with photographs of various types of settlement, included in the 1934 edition of the atlas, were apparently prepared by de Geer. The world population was estimated to have increased from 1,709,700,000 in 1921 to 1,920,000,000 in 1931.

Having noted the vast questions opened up by the work in political geography of de Geer, it is perhaps refreshing to turn to an interesting though wide-sweeping article which he published in 1927 on North America,[3] where he spent several months in 1922. The main argument of his paper is that in America, as in Europe, a single major manufacturing belt had developed. In Europe, a large industrial belt stretches from Britain through northern France and the Low Countries into Germany, Switzerland, north Italy, Bohemia (Czecho-Slovakia) and small adjacent parts of Austria and Poland. Within this area, the population density is generally over 250 to the square mile. Beyond it, there were smaller industrial areas, mainly of local significance and producing goods for a local market: these

[1] See p. 125.
[2] *Nordisk Världatlas*, utgiven av S. Zetterstrand och Karl D. P. Rosen, Stockholm 1926, maps 45 and 46, text pp. 22–7, also Beskrivingar till Kartbladen 37–50: and in the second edition, 1934, maps 45 and 46, text pp. 175–84.
[3] *Geogr. Annlr*, 10, 1928, 205–44.

L

included the Moscow district, Barcelona and its environs, the Bergslagen in Sweden, and the towns, including Copenhagen, on both the Danish and Swedish shores of the Sound. Similarly in the United States, the manufacturing belt extended from the Atlantic Ocean to the Mississippi river and from the Great Lakes on the north to the Ohio river on the south. This belt had

developed during a comparatively short and rather homogeneous historical period and may therefore be expected to have been influenced by geographical laws more manysidedly than has been the case in Europe with its complicated system of states, and its very old traditions.[1]

In fact, the American belt originated on the Atlantic, initially around Massachussetts bay with Boston as a major commercial centre: in time, it grew westwards from Boston, from New York, Philadelphia and Baltimore. De Geer mapped the distribution of wage-earners in cities of 10,000 people and over, and showed that they occurred in groups within several eastern or central states, while most southern and western states had few and rather isolated or dispersed manufacturing cities. Montreal could be regarded as an isolated annexe to the main manufacturing belt, with Toronto as an integral part of it. There were signs of a westward extension, shown especially on the Mississippi, and what might be called 'outposts', developing on the Missouri river between Sioux City and Kansas City.

De Geer considered carefully the location of sixty-six city manufacturing groups, and found that fifty-one of them were based on a river crossing, a canal or a coast line, either of the Atlantic or of the Great Lakes. In short, they had a location favourable to growth before the railway period or the steam age. He considered also the industrial nature of the cities, and notes the intense specialization characteristic of some towns: for example, 39 per cent of the wage-earners in Duluth made Ford cars, and 34 per cent of those in Pittsburgh were steel workers with many more in ancillary industries. But these specialized industrial cities were only of medium size and the very large cities, such as Boston, Providence and Worcester in New England, and above all New York, had a great variety of industries; further, as the specialized manufacturing towns grew they too were acquiring other industries. The area covered by the industrial cities, though little more than 4 per cent of the continent,

[1] *Geogr. Annlr*, 10, 1928, p. 234.

had half its population and in 1920, with 35,600,000 people, had more than two-thirds of its total urban population, then 54,300,000. The shape of the manufacturing belt reflected the history of white settlement: as agricultural activity had pushed westwards, so too had industry, notably in Chicago, though advantages for industry remained strong in the east, largely through the acquired advantages of good communications and power supply.

Although de Geer was aware of current trends, such as the industrial growth in the southern states, he did not foresee some marked recent developments such as the great industrial growth on the western seaboard. His general theories on the concentration of industry within certain favoured areas having advantages acquired through historical growth are still of interest; no doubt many readers have thought of the many discussions in Europe of the need for industry in areas relatively remote from the main lines of commerce and communication. In America, the work of Jean Gottmann on *Megalopolis*[1] shows that great advantages, natural and acquired, still induce commercial and industrial growth in the east, and in Europe the vast growth foreseen by some geographers in the Low Countries and north Germany, in effect the Rhine and its hinterland, underline the same views. On the other hand, the vast Russian expansion of industry covers not only those areas having centrality as an asset, but many others marked by remoteness, and even in Scandinavia some remarkable extensions of industry into areas of sub-arctic climate may be seen. It is possible that the future may see both the concentration and the dispersal of industry, especially if national planning, partly for social reasons, becomes even more marked than at present.

The work of de Geer

In all de Geer wrote ninety-five articles, books, reviews and notes so he was hardly reluctant to put pen to paper. Of all the items in so large a bibliography, probably none was more significant than the population atlas of 1919, which has inspired so many similar efforts all over the world. Even if, as noted above,[2] it was criticized justly for its use of one dot to 100 persons, it gave a clearer impression of

---

[1] *Megalopolis: the Urbanized Eastern Seaboard of the United States*, New York 1961.
[2] See p. 133.

the distribution of population in Sweden than had been available before. Further, it was a notable contribution to cartography widely emulated in the mapping of other distributions, such as the areas of farmed land in Sweden and numerous similar distributions elsewhere.[1] Sten de Geer was not the initiator of the distribution map, but he made it effective without being over-dramatic in its demonstration of the wide disparities in the land use pattern of Sweden; his inclusion of forest limits and other landscape features introduced a measure of correlation that contributed to the regional study of the country. The difficulty, some would say the impossibility, of drawing lines to show distinct areas of Sweden, as of other countries, remains, but at least de Geer's work in this field paved the way for the critics and designers of later, perhaps improved, schemes. De Geer as a writer followed a not-unusual continental pattern by beginning with physical geography and ending with wide human problems. His work on Stockholm, in its time a distinct contribution to the growth of urban geography, showed an appreciation of American ideas then prevalent and gave a basis for the later work on Stockholm by H. W. Ahlmann and W. William-Olsson.[2]

De Geer's work on Sweden is perhaps of more abiding interest than the more wide-flung writings of the later period. His work on the Europe of Versailles was obviously designed to meet a current need and the study of the 'Nordic region' may now seem an academic curiosity, indicative of the enthusiastic adoption by geographers of racial data during the 1920s. Study of the original de Geer article shows that, as the author was aware, the generalizations were made on limited—indeed inadequate—data: the whole basis of racial study is more complicated than many supposed in the 1920s: de Geer, however, made no ascription of mental characteristics to particular racial types. Equally adventurous was his study of past civilizations, in which he showed that he was conversant with the political thinking of Kjellen, normally regarded as the founder of geopolitics, and German workers. Here the main interest lay in the attempt to show

[1] The excellence of the distribution maps in the Atlas of Sweden should be noted, but Sten de Geer was only one of several geographers experimenting with cartographical representation during the period of his active work (and since).

[2] A useful bibliography is given in W. William-Olsson, *Stockholm, Structure and Development*, 1960 (for the International Geographical Congress). See also the same author's paper in *Geogrl Rev.*, 30, 1940, 420–38.

the geographical influences in world history, an aim widely held by geographers of the day. Allowing for all the changes and chances of political circumstance, it was fascinating to trace out, even if tentatively, the geographical factors favourable to the emergence of strong powers that survived for hundreds, even in some cases thousands, of years. But critics would say that this could only be done by generalization to the point of excessive simplification, and that the efforts to trace out a pattern in history were as futile as to explain the rise, continuance and fall of civilizations by geographical factors. Here again, Sten de Geer showed caution, but in the background there was the far more confident Ellsworth Huntington. And though in his general discussion of geography, de Geer spoke of history as valuable for geography on a short-term basis, in this work on past civilizations he showed an interest of a far wider historical range.

Of all the work dealing with large areas, that of de Geer on the American manufacturing belt remains of great value as it is already an historical study showing conditions of 1920 in a country that has changed rapidly since then. The basis of the work was distributional mapping, necessarily of a general character but with a discriminating use of statistical sources. Here, too, de Geer shows that the distribution pattern could only be understood as an expression of the historical evolution of the United States from bases on the eastern seaboard, though this was only one of several explanations, for manufacturing industry could only have evolved with adequate power supplies, natural mineral and other resources, and the provision of good communications. Even so, most of the major industrial centres had sites fixed long before the industrial revolution swept through America, such as river crossings, facilities for ports on the sea, the Great Lakes or canals.

Much of the 'human' and regional geography of the twentieth century, whether studied in a small area or in the whole world, marked a great advance fron the sterile political geography that had preceded it. De Geer looked forward to the creation of a new political geography in which states would be studied as entities, but with a full appreciation of their physical, economic and social geography, to make a new and more satisfying synthesis. The emphasis on the state as an entity was entirely natural to one who had studied the redrawing of the European map with boundaries based on the principle of self-determination, with provision for the rights of minorities: the mapping of distributions such as language and religious affiliations was of direct political significance. In much of his work, de Geer was experimental, looking for new principles and methods of work, new ways of presenting the material. At a time when many broad sweeping generalizations were made by geographers, some of which paid the dividend of attracting at least temporary attention, de Geer was anxious to base his conclusions on adequate data, expressed particularly through distribution maps. Much that he wrote might now seem tentative and only partially true, but he was a man of his time, not content only to study the local minutiae of geography but conscious of the possible contribution that could be made to citizenship, not only of Sweden but of Europe and the world. For his time, he showed much caution and restraint: he was a very reasonable man.

BIBLIOGRAPHICAL REFERENCES

The obituaries of Sten de Geer include a careful study by J. Leighly in *Geogrl Rev.*, 23, 1933, 685–6, and a short appreciation by C. B. Fawcett in *Geography*, 18, 1933, 230. There is a fuller treatment by Helge Nelson in *Svensk Geografiska Årsbok*, Lund, 1933, 185–92, with an English summary. *Gothia*, Göteborg, 3, 1934, 1–21, 23–8, has a life by Sven Swedberg and a bibliography compiled by K. G. Tengstrand. H. W:son Ahlmann wrote a study of de Geer in *Ymer*, 53, 1933, 441–5.

Several of de Geer's papers were published in English, French or German, at least in summary form. His publications fall into groups:

1. *Physical geography*
The earliest are 'Dalälfen och dess utskärnigar nedom Alfkarlebyfallen', *Ymer*, 26, 1906, 83–92, and 'Om Klarälfen och des dalgång', *Ymer*, 26, 1906, 383–414. Then followed 'Niplandskap vid Dalälven' in *Sveriges Geologiska Undersökning, Årsbok*, 9, 1912. The work on harbours is in 'Storstäderna vid Ostersjön', *Ymer*, 32, 1912, 41–87, which appeared in a shortened form as 'Die Grossstädte an der Ostsee', *Zeitschrift der Gesellschaft für Erdkunde zu Berlin*, 1912, 754–66 (no volume number). Years later, with some newer material, it was developed as *Svenska Hamnförbundet Östersjöhamnar nas Geografi*, Stockholm 1927. Lake analysis is the theme of 'Geografisk undersökning av sjöarna Toften, Testen och Tysslingen i Närke' in *Sveriges Geologiska Undersökning, Årsbok*, 4, 1912 and other publications in the same *Sveriges Geologiska Undersökning*, 1910 and 1913. The short study of a glacier in Spitsbergen appeared in *Ymer*, 33, 1913, 148–57.

2. Population studies

Of these the first publication was 'Befolkningens fördelning på Gottland', *Ymer*, 28, 1908, 240–53, and the title of the atlas is *Karta over Befolkningens Fördelning i Sverige den 1 Januari, 1917*, Stockholm 1919, with a text volume in Swedish. Information in English is given in *Geogrl Rev.*, 12, 1922, 72–83. A short article on the same theme appeared in *La Géographie*, Paris 37, 1921, 517–24.

3. Regional and urban geography

The paper on the political regions of Scandinavia appeared in *Ymer*, 1918, 24–38. The major human regions were discussed in 'Om Sveriges geografiska regioner', *Ymer*, 45, 1925, 392–415; a shorter version of this paper appeared in *Erde und Wirtschaft*, 2, 1927, 54–67. A paper on north Sweden's regionalization is 'Norra Sveriges Land-formsregioner', *Geogr. Annlr*, 8, 1926, 125–36, and the delimitation of Fennoscandia is in German, 'Das geologische und das geographische Baltoskandia', *Geogr. Annlr*, 10, 1928, 120–39. The work on Stock-holm originally appeared as *Storstaden Stockholm ur geografisk syn-punkt*, Stockholm 1922, and later in a shorter English version in *Geogrl Rev.*, 13, 1923, 497–506. De Geer's study of industry in the United States was first published in English as 'The American manu-facturing belt', *Geogr. Annlr*, 9, 1927, 233–359. Later it was published in a shorter Swedish form and used for an inaugural address at Göteborg.

4. Political geography

'Europas statsgränser och statsområden efter varldskriget', *Ymer*, 40, 1920, 253–302, was developed into *Det Nya Europa*, Stockholm 1922. 'The kernel area of the Nordic race' appeared in *The Racial Characters of the Swedish Nation*, Uppsala 1926, for which see the footnote on p. 144. The main historical paper is 'The subtropical belt of old Empires', *Geogr. Annlr*, 10, 1928, 205–44. The paper on the content of geography appears as 'On the definition, method and classification of geography', *Geogr. Annlr*, 5, 1923, 1–37.

CHAPTER 7

Two Modern British Geographers

O F the various modern geographers in Britain not now alive, perhaps the most widely known was Halford John Mackinder (1861–1947), who became Reader in Geography at Oxford in 1887. He was the first of many teachers appointed in the universities from this time forth to develop the teaching of geography but he was by no means the first teacher of geography in British universities, for he himself said that the first Reader at Oxford was the great Elizabethan geographer Richard Hakluyt (1552–1616), and there were intermittent periods when geography was taught in colleges of the University of London before 1887. But from this time there was continuity of development in the universities of Britain and this chapter deals with two geographers who belonged to this modern phase of development: the life and work of Mackinder has been discussed in various articles reasonably accessible to those who wish to know more of him.[1] In this chapter attention is given to P. M. Roxby (1880–1947) and A. G. Ogilvie (1887–1954), two geographers whose main work belonged to the inter-war period and whose interesting though contrasting experience and personalities brought them deep respect from a wide circle of people extending far beyond the academic enclaves of the cities in which they lived. Not for one moment it is claimed that they were greater than others and both of them would have been surprised—even embarrassed— by their inclusion in such a book as this.[2]

PERCY MAUDE ROXBY

Roxby was one of those people who—like a character in a novel by Jane Austen or Ivy Compton-Burnett—reveal their character in

[1] Notably in Gilbert, E. W., obituary in *Geogr. J.*, 110, 1947, 94–9; Unstead, J. F., 'H. J. Mackinder and the new geography', *Geogr. J.*, 113, 1949, 47–57; Gilbert, E. W., 'Seven lamps of Geography', *Geography*, 37, 1951, 21–43; *Sir Halford Mackinder 1861–1947*, Centenary Lecture, London 1961, and in the *Dictionary of National Biography*.

[2] Both were well known to the present author: some of the material used here has been drawn from personal letters and conversations.

PLATE 4. P. M. Roxby

almost every word and gesture. He had an immediate impact on an audience, not least because he was an impressive figure, 6 ft. 5 in. in height, with a strikingly rugged face and a propensity to dramatic emphasis of each point as it arose. In short, he had *panache*. Every word of the lecture was prepared with meticulous care, and as each slide was shown the various points of detail were enunciated one by one. Essentially the lecture was a dramatic performance and he was a great believer in the spoken word as a means of communication: for an academic, he had a singularly unselfish outlook as he would attend innumerable meetings, speak to obscure societies and give lavishly of his time to such enterprises as courses for the Workers' Educational Association. All this was in some sense part of his educational responsibility, but his natural generosity may have prevented him from realizing that a university teacher may reasonably keep some time for himself to read and write. With truth he could say, 'I never have a moment during term': he had not. If he had to lecture, he wanted at least half-an-hour of complete silence before it to go through his material—not to prepare it, for that had been done long before: if some student wanted an interview, he would allocate plenty of time to him, especially if the student was Chinese or Egyptian: if some citizen of Liverpool wanted him to assist with some good cause, he would be graciously received and promised help in one form or another. He was an ardent supporter of good causes, and thought it perfectly reasonable that a junior colleague should give half his time to work for the League of Coloured Peoples: whether the university as a whole thought so, he never troubled to enquire. To any young geographers of a later time, such an attitude may seem strange, but there were two characteristics of his time that were partly instrumental in this approach. First, it was clear that geographers must go out to the world and present their views, even to small groups and societies, and second, in some universities, not least Liverpool, the link with the city was so strong that civic service in one form or another was a natural outlet for the public-spirited.[1]

By training, Roxby was an historian. His early years, spent at

[1] Since the 1939–45 war the universities have become in effect national institutions. No longer need they draw their students mainly from the area in which they are situated: but many geography departments still preserve the good tradition of study of the city and region in which they are placed.

Buckden, near Huntingdon, where his father was the rector, gave him a permanent love of the countryside combined with a great interest in old churches. One of his students complained that Roxby's idea of field work in East Anglia seemed to consist of a kind of 'church crawl' from one village to the next: his intimate knowledge of East Anglia, acquired in leisurely visits through many years, was shown in a regional study published in 1928.[1] Educated at Bromsgrove School, Roxby went to Christ Church, Oxford, and graduated with first class honours in history in 1903. As an undergraduate, in 1902, he had won the Gladstone memorial prize for an essay on Henry Grattan. But even before he graduated Roxby went to lectures in geography under A. J. Herbertson (1865–1915) and H. J. Mackinder who had then (1902) recently published *Britain and the British Seas*, long regarded as a classic, followed by 'The geographical pivot of history',[2] which at the time and later was widely discussed as a contribution to world political geography. Within a short time after Roxby graduated Liverpool University, newly emerged from the federated Victoria University that also included the colleges at Manchester, Leeds and Sheffield, wished to begin the teaching of geography, and Roxby was asked to do this under the friendly guidance of an economist, Professor E. C. K. Gonner, who advised him that the best introduction to Liverpool life was to spend several afternoons at the docks. And this he did. Roxby was in such a hurry to get to Liverpool that Sir John Myres spoke of him as 'snatched away half-trained' and noted that he had to return to Oxford to take examinations in geography.[3] In fact Roxby was largely self-taught as he had neither the time to profit from the teaching in Oxford nor the opportunity of studying in continental universities, where several famous geographers were working. Though a place so different from everything he had previously known, Roxby settled happily in Liverpool, partly because he found it interesting historically and partly because of its fascinating human problems, having immigrant communities in large numbers. Permanently aware of its ugliness, he watched with interest the growth of the university and rejoiced in the co-operation with the city that became its avowed policy.

[1] See pp. 163–6.

[2] *Geogr. J.*, 23, 1904, 421–37.

[3] Private communication. Incidentally, Roxby never bothered to take his M.A. in Oxford and had no use for degree-hunting.

The work on China

Crucial to Roxby's experience as a geographer was the year he spent as Albert Kahn fellow in 1912–13, during which he visited the United States, India and China. It was the last of these areas that attracted his interest most, and on which he wrote a limited number of papers and, during the 1939–45 war, part of the Admiralty Handbooks on China. His first article on China[1] was a study of a possible capital, for which the trio of cities, Hanchow, Wuchang and Hanyang, in the central basin of the Yangtze appeared to be most appropriate because they were in the heart of a country acquiring railways and motor roads. But the article which brought Roxby far more readers was the summary of the China Continuation Committee's book, *The Christian Occupation of China*, published in 1922.[2] This work, based largely on the observations of missionaries and others who had penetrated various parts of China, gave population figures for the various provinces along with much detailed information, and in the circumstances ranked as a main source. Roxby held the view that the real cohesion of China lay in the prevalence of the Confucian ethic, expressed particularly through the unity of the family structure and the village with a system of central government based on the nominally absolute power of the emperor supported by a bureaucracy recruited by examinations in the Chinese classics. Equally he was fascinated by the long history of China, and saw its growth as a vast territorial entity partly as the outcome of its occasional disintegration into warring but ephemeral states ultimately re-united into one whole by some powerful ruler, generally in origin a barbarian from the steppelands of Mongolia or Manchuria.

In later articles, Roxby faced the question of the identity of China. 'Is it an entity in any real sense?' he asks.[3] 'If so, can it hold together under obviously increasing pressure?' He stated that 'the study of its human geography may not provide a direct answer to the questions of whether and how China will attain political unity but it can at least make clearer the setting of the problem and some of the factors which are likely to decide the issue'. By some writers, Roxby noted,

[1] 'Wu Han, the heart of China', *Scott. Geogr. Mag.*, 32, 1916, 266–78.
[2] 'The distribution of population in China', *Geogrl Rev.*, 15, 1925, 1–24.
[3] 'China as an entity: the comparison with Europe', *Geography*, 19, 1934, 1–20; 'The expansion of China', *Scott. Geogr. Mag.*, 46, 1930, 65–80.

China had been compared with Europe as a whole, but though this was a useful guide to those who wanted some idea of its size it was little else, as China had seen a long-sustained, almost continuous expansion of a distinctive type of civilization from a single area of characterization, in the valleys of the loess plateau in the northwest and the adjacent drier portions of the Northern Plain round the middle and lower courses of the Yellow river. In a later paper,[1] Roxby criticizes Arnold Toynbee's statement[2] that the emergence of Chinese civilization in the Yellow river basin was an example of 'the stimulus of land environments' (because) 'ease is inimical to civilization . . . the stimulus towards civilization grows stronger in proportion as the environment grows more difficult'. After a review of the evidence available on the early civilization of China, Roxby reaches the conclusion that

. . . the essential geographical element in the rise of early Chinese civilization would seem to have been the existence of an almost continuous East–West belt of relatively forest-free and fertile loess soil, initially favourable in spite of some handicaps, which admittedly may have acted as a spur to agricultural development, and also open on its continental side to the entry of fresh cultural stimulus from Western Asia.

The fascinating story of the spread of Chinese civilization is told in the 1934 article, which includes maps by a Chinese research worker showing the distribution of population according to two early censuses, of A.D. 2 and A.D. 140.

Roxby's final work on China was done under conditions of stress in the editing and preparation of the three-volume *Geographical Handbook of China* published by the Naval Intelligence Division[3] and released for general sale some ten years after publication. It would not be fair for the present author to indicate which parts of these volumes were in fact written by P. M. Roxby as the work was a co-operative enterprise. At least it gave a stimulus that induced him, protesting vigorously at the 'speed of production', to write down much that he knew on China but had never committed to the printed

---

[1] 'The terrain of early Chinese civilisation', *Geography*, 23, 1938, 225–36.

[2] *A Study of History*, vol. 1, 318–21; vol. 2, 31–3. Roxby's general admiration for Toynbee's work was considerable.

[3] *China Proper*: vol. 1, *Physical Geography, History and Peoples*, 1944; vol. 2, *Modern History and Administration*, 1945; vol. 3, *Economic Geography, Ports and Communications*, 1945.

page. Through the co-operation of colleagues he was spared the worry of arranging the illustrations, for he regarded illustrations as an addition to the text rather than an essential part of it: he had the artist's appreciation of the value of a word and once told the present writer, 'I often have to walk round Grantchester Meadows for three-quarters of an hour to find the right word.' In fact the technique of illustration has advanced markedly since Roxby's main work was done, and to some extent his attitudes were due to his training as an historian. This was perhaps particularly so in his attitude to printed sources. He regarded some, such as the *Christian Occupation of China* volume, as primary sources worthy of respect almost amounting to veneration, and was equally enthusiastic over some books and articles.[1] For an academic, he read deeply rather than widely, and studied texts extremely closely, analysing the meaning of each sentence. He was in many ways a perfectionist in his own writing and expected to find the same austere standards in others.[2]

Regional and human geography

An early article of Roxby[3] began as a study of William Marshall (1745–1818) who made a thorough survey of the rural economy of England between 1787 and 1798 and was partly responsible for the foundation of the Board of Agriculture in 1793. Roxby quotes Marshall's statement that

Natural not *fortuitous* lines are requisite to be traced; *agricultural* not *political* distinctions to be regarded.
A *Natural District* is marked by a uniformity or similarity of soil and surface, whether by such uniformity a marsh, a vale, an extent of upland, a range of chalky heights, or a stretch of mountains be produced.

Marshall notes that an agricultural district shows uniformity of farming in grazing, sheep farming, arable management or mixed cultivation, and probably some particular speciality such as dairy

[1] The author is not suggesting that historians take the printed word as gospel truth: quite obviously Marc Bloch did not. Roxby undervalued map sources and so presents an interesting contrast to Ogilvie, whose use of maps is discussed on pp. 169–70.

[2] There were times when the present writer could only stem his protests at the inevitable 'speed of production' of the *Handbooks* by holding out the prospect of more leisurely composition after the War.

[3] 'The agricultural geography of England on a regional basis', *Geogr. Teach.*, 7, 1913–14, 316–21.

produce or 'fruit-liquor', presumably cider and perry. Such agricultural districts did not correspond to counties: it was, for example, more logical according to Marshall to divide the Severn valley into the Vale of Gloucester and the Vale of Berkeley, with the Vale of Evesham as a continuation of the former, than to study the agriculture of counties: far better results would be gained by noting that the dairy district of north Wiltshire included parts of Gloucestershire and Berkshire and also the eastern margins of Somerset. Roxby noted that no subsequent agricultural economist had followed this regional method, and argued that a regional division on the basis of relief, geological formation and climate could be an effective basis for the study of agricultural distributions, including crops, pastures and stock, and that the population statistics could be similarly studied for each 'natural district'.

The idea of natural regions was developed further in later work.[1] An article of the middle 1920s reviews the use of regions among a variety of geographers working on different scales. Obvious political examples included the case of Austria and Hungary in relation to the Danube basin: incidentally, though Roxby never published his views on the treaties that followed the 1914–18 war, he held the view that Austria was too small to be an economic unit and that its union with Germany in 1938 was unavoidable and probably finally justifiable: he was also of the opinion that the inter-war Poland was too large and that the inclusion of such considerable minority groups as the White Russians and the Ukrainians was dangerous. Such views he shared with many people in the universities at the time. More specifically, it was recognized that there were many types of regions, for example those in the *Oxford Survey of the British Empire*, though called 'natural', were in fact physiographic and why not therefore say so? Equally much of the writing on the United States was linked to a widely known scheme of physiographic regions, though 'human use' was the basis of the equally well-known 'corn and winter wheat belt' and the other agricultural division of the United States. Economic orientation offered another interesting basis, seen especially in the work of C. B. Fawcett's[2] division of England into areas tributary to a number of regional capitals. Recognizing the value of the

[1] 'The theory of natural regions', *Geogr. Teach.*, 13, 1925–6, 376–82.
[2] *The Provinces of England*, London 1919—revised with a new introduction by W. G. East and S. W. Wooldridge, London 1960.

regional scheme of A. J. Herbertson,[1] Roxby noted that the climatic basis might be inadequate, as the 'natural forests' might not be the same in regions climatically similar. Equally, no allowance was made for 'space relations', a term highly characteristic of the Roxby teaching. Although the Sahara and the Atacama deserts, the Mediterranean basin, central California and southwest Australia might have comparable climates, their human position in the world was vastly different. The division of Europe on a climatic basis was useful but not adequate in itself, for one of the continent's most striking features was the parallelism of the three main structural belts, the Northern Lowlands, the Central Uplands and the Alpine areas. Central Europe in Roxby's view was definitely part of the European peninsula, and from prehistoric times onwards had shared in the cultural advantages which the interchange of trade between the Mediterranean and the 'Northland' conferred, but it was much less favoured than oceanic Europe as it experienced some isolation due to the enclosure given by the Alpine barrier and the minor significance of the Baltic sea compared with the narrow seas of western Europe. Another adverse factor, historically, was its exposure on the east side to invasions from the vast heartland plains of the Old World, from which western Europe was largely immune. In such an approach one can recognize the broad sweep of ideas that appealed to or annoyed the hearer or reader according to his views. But the article also commends the use of the small entity, in fact the *pays* of the French regional geographers: and modern agricultural specialization, instead of decreasing the influence of physical conditions on farming, in fact made them all the stronger, in Roxby's view for 'the norm of regional specialization is use adapted to *intrinsic* conditions'.

And this theory was worked out strongly in the essay on East Anglia in the 1928 volume of regional essays published on the occasion of the International Geographical Congress in Cambridge.[2] In this essay, East Anglia is defined as coincident with the ancient kingdom of the same name, delimited by the Fens, and the great Essex forest on the London clay, but less isolated on the southeast and the southwest. After a survey of the physical geography, and particularly of the glacial deposits, Roxby discussed the main features

[1] 'The major natural regions of the world', Geogr. J., 25, 1905, 300–10.
[2] Ogilvie, A. G. (ed.), *Great Britain: Essays in Regional Geography*, Cambridge 1928, 143–66.

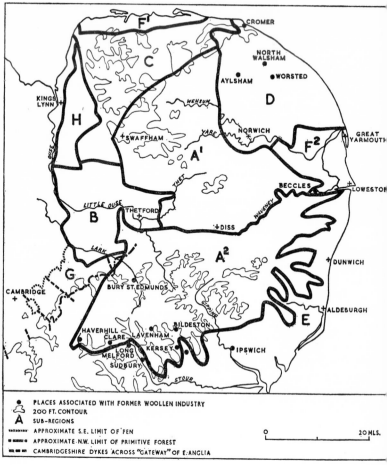

FIG. 19. A regionalization of East Anglia

As explained in the text, P. M. Roxby divided East Anglia into sub-regions of the *pays* type. The sub-regions are:

A[1] *High Norfolk.* Mainly heavy and tenacious boulder clays, mixed arable farming: many commons.

A[2] *High Suffolk.* Similar to above but heavier soils, especially in the south.

B *Breckland.* 'Blowing sand' on chalk: much uncultivated heath. Various agricultural experiments tried here: now large forest areas.

C *'Good Sands'.* Not initially attractive but brought into cultivation.

of the climate and 'natural vegetation'. This last is a subject most geographers would approach with diffidence forty years later and his main point was that certain areas were either 'open heath or grassland' and inviting to prehistoric settlers. This leads to a section on 'historical and human' geography. in which the choice of various sites and areas for settlement at different historical periods is noted: for example in early Norman times East Anglia was 'the most populous region in England'. True to his experience and tastes, Roxby notes that 'confirmatory evidence is afforded by the closeness of the nucleated villages to each other and the relatively immense number of parish churches'. The trade in wool with Flanders is discussed, and the cognate industrial growth of the fourteenth to the eighteenth centuries noted: we are even informed that 'Mr Havelock Ellis, in his study of the distribution of British genius, gives the East Anglian focus first place.' Vastly significant to the area was the Agrarian Revolution, followed by a limited industrial expansion. The limitation of industry was in fact an advantage for

. . . the well-tilled fields painted by Constable, Crome, and many other artists of a notable school, have escaped the devastating phases of modern industrialism. Few countrysides are richer in historic associations or in the evidence of human achievement over long centuries. There is no sharp separation between rural and urban life. Industry is linked with agriculture and many a beautiful country town like Bury St Edmunds has developed modern industries without losing the evidences of its antiquity.[1]

[1] Ibid., 166.

D   *'Loam Region'*. Light but fertile and easily worked glacial loams. Long occupied, rich farming with roots, barley, wheat and especially bullocks.

E   *East Suffolk*. The 'sandlands' or 'sandlings' are light glacial loams, sand and gravel. Some fertile areas, but heathland also.

F[1] *North Norfolk Marshland*.

F[2] *Broadlands*. Valuable alluvial land, summer grazing for farms in C and D.

G   *Chalk Downland*. Formerly dry upland pasture but converted to farmlands with wheat, barley and sheep.

H   *Greensand Belt*. Broken hilly country, wooded ridges and heaths, some fertile valleys with relatively good pasture.

Originally published in Ogilvie, A. G. (ed.), *Great Britain: Essays in Regional Geography*, Cambridge 1928, 164–5

The culmination of the essay is a division into nine sub-regions, each having a distinctive type of soil and farming: of these it is noted that

they may be considered as comparable in content and character to the well-known *pays* of the Basin of Paris, although lacking an equal heritage of historical associations. They do, however, indicate geographical realities and may be treated as entities for the study of settlements and of agricultural developments.

The imprint of Vidal de la Blache's teaching in the work on East Anglia will be clear to readers of chapter 3. And in a 1930 paper[1] Roxby traces the modern growth of geography back to the work of Ritter and Humboldt, but states that there was a 'modern geography' only in the sense that there had been 'a restatement of its scope and content in the light of all the new knowledge of the earth which more specialized branches of enquiry have revealed'. There was constant change in the world, so that the stage on which man's activities were set was itself changed from one time to another: Roxby could not agree with Ratzel's view that the stage was always the same and always situated at the same point in space, nor could he accept Demolins' idea that if it all began again, history would have the same course. It was within human power to look forward, as Ritter had said with considerable eloquence, and this could be done by regional planning, which was

essentially a conscious effort in constructive social geography, the attempt to utilise all elements in the physical environment for social well-being (as distinct from the ruthless exploitation of particular elements, e.g. coal, regardless of the wider social consequences, which marked the earlier stages of the industrial revolution) and to harmonise the interests of neighbouring towns and countrysides in a common scheme in which each has its place.

Human geography was concerned with the adjustment of human groups to their physical environment, including the analysis of their regional experience: it also involved the study of inter-regional relations as conditioned by the several adjustments and the geographical orientation of the groups living within their respective regions. Just as it was necessary to find the right relationship between geography and geology, so too it was necessary to find the right

[1] 'The scope and aims of human geography', *Scott. Geogr. Mag.*, 46, 1930, 276–99. This was the Presidential Address to Section E of the British Association.

relationship between human geography and anthropology, for the geographer must consider not only the natural resources and economic possibilities of a region but also the adaptability, climatic or otherwise, of various racial groups for developing them.[1] Economic geography could be studied for its own sake, but should be related also to social geography, which Roxby defined as 'the regional distribution and inter-relation of different forms of social organization arising out of particular modes of life': some of these, such as nomadic regimes, were directly related to the physical environment. But changes were possible: Denmark provided an example of effective co-operative effort by which a country was made infinitely more prosperous than it had been before. On political geography, Roxby notes that no modern development had made it less important; the 'New Europe' of the Versailles and later treaties was itself 'a great experiment in political geography'. All of these could be treated historically, for historical geography was 'essentially human geography in its evolutionary aspects', concerned with the 'relation of human groups to their physical environment and with the development of inter-regional relations as conditioned by geographical circumstances'. Historical geography demanded 'the reconstruction of the physical stage in different phases of development'; that is, it gave a picture of man and environment at any period in history which might be studied.

All through the work of Roxby his deep historical sense was apparent. His writing shows breadth rather than depth, though a careful reading of the East Anglia article will show how thoroughly he knew and loved his native countryside. He arranged field tours for his students, but under the leadership of local experts, as like many geographers of his time, he was not at ease in this type of work. Generally he saw landscapes largely as the areas where people lived, and his writing on the physical aspects is less vivid than that on the human aspects. In 1944 he retired from the Chair of Geography at Liverpool which he had held from its foundation in 1917, and went to live in his native Buckden. In the following year, with his wife, whom he married in 1941, he went to China as chief representative of

[1] Since 1930, the whole attitude to the inclusion of racial material has changed, partly because of its flagrant misuse in Germany and partly because many of the initial assumptions made on both the physical and the mental qualities have been largely disproved by later research. But see pp. 142–6. Roxby was thinking mainly of physical anthropology.

the British Council. Letters to his friends show that he was perplexed by the mounting political difficulties of the time and harassed by administrative responsibilities. For years he had believed that Britain had given inadequate help to China, especially in educational enterprise, and he conceived his work as a contribution to its rehabilitation. In 1946 he died suddenly in China.

ALAN GRANT OGILVIE

A. G. Ogilvie was the only child of Sir Francis and Lady Grant Ogilvie, Principal of the Heriot-Watt College in Edinburgh and later at the Royal Scottish Museum and at the South Kensington Museums. Born in Edinburgh in 1887, he was educated mainly in England at Westminster, and at Magdalen College, Oxford, where he graduated in history in 1909: at that time no degree was available in geography. He had, however, already met A. J. Herbertson, who was working in Edinburgh for the Bartholomew map firm. Ogilvie became a post-graduate student at Berlin University, where he met the great Albrecht Penck, then at the height of his powers. In Paris at the Sorbonne he encountered the famed French geographers of the time, Vidal de la Blache and Emmanuel de Martonne. He also met W. M. Davis and attended his lecture course at the Sorbonne. He was a member of the transcontinental Excursion across America in 1912, on which he came to know some eighty geographers of varying fame and ability. It was a liberal and in some ways a leisurely introduction to his life work and in February 1912 he became a junior demonstrator in Oxford on the resignation of O. G. S. Crawford, since widely known as an archaeologist. In Oxford, he gave lectures on general and physical geography, assisted with a course on the British Isles, organized the library and helped with field work, much of it done on bicycle tours of the neighbouring countryside. On one occasion he was alarmed by a violent crash due he thought to the fact that a woman student who was a militant suffragette had broken another window: fortunately it was not so.

While resident in Paris, Ogilvie collected a great deal of material on Morocco in the Bibliothèque Nationale, which he presented as a paper to the Research Division of the Royal Geographical Society in 1912. Warmly commended at the time for this work,[1] Ogilvie had

---

[1] 'Morocco and its future', *Geogr. J.*, 39, 1912, 554–75: 'Notes on Moroccan geography', ibid., 41, 1913, 230–9.

not then visited Morocco but took the view that through map work
and various sources one could construct an image of a country.
Though he was humorously apologetic about this early work in his
mature years, he was firmly of the opinion that intelligent map study
can be a substitute for field work giving at least partial satisfaction:
in any case it is hardly possible for a writer, even on a small country,
to see every corner of it. The right use of the map was something he
carefully considered in all his teaching: in one Edinburgh meeting
of the university staff, a distinguished ancient historian spoke of
the difficulty of teaching students to discern all there was in any
historic manuscript. How similar, Ogilvie commented afterwards,
was the problem of teaching students to see all there was in a map
and to analyse it effectively. At the same time, he did not favour the
'guess work' sometimes advocated in textbooks. He was deeply interested in the teaching of map work to the Pass students of the university, then numerous, and deeply critical of many of the text-books
(and their writers) filled with shallow generalizations. His Honours
students were taught regional geography, which he placed in the final
year as the culmination of the course, with the full use of detailed
topographical maps and visual aids such as photographs and slides.

Ogilvie was to have toured the world in 1914–15 as an Albert
Kahn fellow, two years after P. M. Roxby,[1] but this was not to be.
He served in the Royal Field Artillery and in a Field Service Unit,
and in 1918 was sent to the Geographical Section of the General
Staff, with which he went to Versailles as a member of the British
Delegation. Part of his work was to make a survey of southern
Macedonia, and in the Balkans he acquired a great deal of experience
which was to be the basis of valuable contributions to the Admiralty
Handbooks on Greece published nearly thirty years later. He conceived the task of the map makers at the Versailles conference as
the delineation of boundaries that would be least likely to make
trouble afterwards, but he was fully aware of the dangers that lay
ahead and as early as 1935 regarded the possibilities of European
war as grave: a visit he paid to Germany in that year did not lessen
his fears. The principles underlying the territorial decisions of the
Peace Conference[2] were partially geographical, due largely to the

[1] See p. 159.
[2] *Some Aspects of Boundary Settlement at the Peace Conference*, No. 49 of
'Helps for students of history', London 1922.

cohesive work of the American 'Inquiry' centred at the house of the Geographical Society in New York: the British government drew its data from a number of different departments, with much overlapping. In seeking to change the political allegiance of people from one state to another, the Conference tried to reduce ethnic variety and also to leave the new or altered states with adequate communications, both for ordinary commercial movement and for military purposes. The 1 : 1,000,000 map was the standard base for boundary delimitation with more detailed work on the 1 : 200,000 maps or those of similar scales. The military value of hill ranges, the military and economic value of railway junctions, and the retention within one state of natural entities such as intermontane basins and valleys were considered. Inevitably the politicians were able to overrule the experts and some of the frontiers were not drawn up on principles that could be called geographical. In fact, many boundaries included large minority groups, such as the Magyars, both in western Rumania, where the boundary through the Danubian plain was so drawn to include a railway, and in the Slovak part of Czecho-Slovakia, which was provided with good railway links and a foothold on 100 miles of the Danube. In the pamphlet of 1922, Ogilvie considers critically the boundaries of Czecho-Slovakia, Yugoslavia, Rumania and Bulgaria with Italy, Austria and Hungary as defined in the treaties of Versailles, St Germain, Trianon and Neuilly. His work shows a deep knowledge of military strategy and also of the fundamental regional geography of the time.

In 1919 Ogilvie went to Manchester University, but left after one year to join the staff of the American Geographical Society, where one of his main tasks was to write *The Geography of the Central Andes*, published in 1922 as one of the Society's Research series. This work was a contribution to the general survey of Hispanic America, and whose partial culmination came in the publication of a complete series of 1 : 1,000,000 maps for which the Society had assumed responsibility.[1] In fact, Ogilvie never went to the Andes, and was entirely dependent on maps, photographs and printed sources, so the gift for evoking landscapes was fully tested in this

[1] An account of this enterprise is given in Wright, J. K., *Geography in the Making, The American Geographical Society, 1851–1951*, New York 1952, 300–19. On p. 304 it is noted that each sheet was to be accompanied by a handbook, but Ogilvie's work was the only one published.

work. In 1923, he returned to Edinburgh, to follow G. G. Chisholm, the economic geographer, and there he remained to his death in 1954. The work at Edinburgh was begun in his middle thirties with a rich and varied experience marred by intermittent ill-health dating from his war service. He came to a city he knew well and where he was known, and one with an established geographical tradition centred around its map firms (especially Bartholomew's), the work of Chisholm in the university and the successful career of the Royal Scottish Geographical Society with its distinguished Editor, Dr Marion I. Newbigin. As the years went on he was drawn into a wide range of public activities, too many for his strength and certainly too many to permit the enjoyment of long clear periods of study and writing. Like many other unselfish people, he was almost oblivious of the social contribution he made to his day and age through his work for so many societies and organizations; and he was always advising others to save some time for recreation and necessary rest. But though he regarded his published output as slight, his work shows some interesting features and, it will be noted, an outlook quite different from that of his near-contemporary, P. M. Roxby.

The physical basis

At all times, Ogilvie insisted that geographical study is incomplete without an adequate study of the physical basis, and this was most clearly stated in a paper given as the Herbertson Memorial lecture[1] in 1937 at Newcastle-on-Tyne. Referring to the work of Herbertson, he notes that the major 'world regions' which he and later geographers defined were climatic and suggested that these large units could be conveniently subdivided on a land forms basis. If one agrees that 'the character and distributions of all phenomena studied by geographers are intimately bound up with the relief of the land' then to give a satisfactory regional view 'the thoroughness of the geographer's study of the land forms ought to be in direct proportion to the degree of detail in which he views and discusses the other phenomena rooted in a region'. Though physical geography is the last chapter of the geological story, few geologists in Britain have shown much interest in geomorphology and most geomorphologists

[1] 'The relations of geology and geography', *Geography*, 23, 1938, 75–82.

have worked in schools of geography[1] and have been concerned with the evolutionary and explanatory description of land forms rather than with a merely empirical approach. Ogilvie in fact was much concerned with the evidences of peneplanes or surfaces of erosion in the landscape and wished students to know how such features had evolved. Yet he agreed wholeheartedly with W. M. Davis in his view that the geographer, as such, should begin with the results of geomorphological work re-written for his own purposes, for '. . . the geographer is concerned with the present, which he describes with the main verbs in the present tense'. In fact the geographer 'may well find real use for calculation from the map of such things as stream density, areas of undissected remnants of a given erosional or constructional slope, average degree of slope, stream length compared to valley length, proportion of slopes of given exposure . . .'[2] In the same paper Ogilvie asks that such phrases as 'old hard rocks' should be qualified as, for example, 'the (resistant) Cambrian quartzites of western Britain are represented in northern Russia by weak clays of the same age'. And such phrases as 'young fold mountains' are equally misleading as in fact many mountains of the world owe their present form to recent uplift followed by erosion by water and possibly ice also. The theories on the use of physical geography put forward in the paper now under discussion are borne out in the essay on Central Scotland,[3] in the 1928 volume, where the existence of surfaces at different levels within the lowland between the coast and the Moorfoot hills is stressed, along with the raised beaches near the Firth of Forth.[4] Other features of this lowland include a system of glacial meltwater channels at right angles to the present main streams, and the deep incision of the major rivers within the surfaces or erosion. Many other features of the physical geography were noted in this essay, including the far higher stream density of the west compared with the east. The real point is that one saw all this in the landscape as one walked and rode across it.

But what is the explanation of such land forms? Their recognition on the surface, or by various methods evolved by geomorphologists such as the construction of projected profiles, is only a first stage.

[1] The American position has been different, for the geomorphologists generally work in Geology Departments.

[2] *Geography*, 23, 1938, 77.

[3] In *Great Britain, Essays in Regional Geography*, Cambridge 1928.

[4] See pp. 175–8.

For the regional geographer, the land forms are of great significance: though he is not, as a geographer, concerned with tectonic history, a full description of the structure may find a place in the geographer's 'first chapter', but only in certain circumstances. The complex folding and overthrusting in the Scottish Highlands may have little influence on the surface forms, but the effects of later dislocations may be considerable; for example, the Great Glen fault is seen as a perceptible landscape feature, and Inverness is still a seismic epicentre. The profound structural break along the edge of the Himalayas is clearly one of the great geographical features of the Asian continent, and obviously related to the tremendous gorges through the mountains and to the occasional severe earthquakes. And in the Wasatch mountains of North America the fault scarp is so recent that the terminal moraines at its foot exhibit miniature rift valleys.

Although some geomorphological processes are visible by direct observation or by the clear imprint of some recent movement, much can only be discerned by inference from observation in the landscape. In the teaching of geomorphology, Ogilvie generally followed the cyclic theory of W. M. Davis, and devised means for illustrating this experimentally in an earth sculpture laboratory.[1] This consisted of an oblong box 8 ft long, 5 ft wide and 1 ft deep, in which silt was placed, with a 'wave machine' to simulate the reality, a small fine spray in the ceiling for rain and piped water for rivers. There were various minor and some major complications in the initial stages. At first the 'rain' abruptly ceased on occasions, due to the fact that the water supply came through the same pipe that fed a cistern in the women's cloakroom next door. The tank itself could be placed at any desired angle to represent the varied slopes of initial surfaces. Such mechanical aids to the teaching of landforms are now found in many universities: on a simple level, their value lies in the demonstration of the three-dimensional earth to students. Ogilvie claimed that they could simulate many forms seen in nature, and especially those of shoreline development, as in this case neither inevitable and permanent saturation of the material (silt) nor the excessive strength of the waves provided by the machine really mattered. By lowering the water level, a raised beach could be simulated, and then subjected to further landform sculpture such as weathering and river erosion. Along with this work, Ogilvie used numerous block diagrams in his

[1] 'The earth sculpture laboratory', *Geogr. J.*, 87, 1936, 145–9.

teaching in the effort to make physical geography comprehensible and attractive to his students. In this he was facing the problem that confronts every teacher of this subject—how to visualize a landscape intelligently from maps, descriptions or even when actually there. It is a far more difficult art than many people appreciate as the power of discriminating observation is developed only by experience. Special care was given to the large classes taking the Pass degree course, many of whom regarded all learning as something enshrined in books and nowhere else and had apparently been taught by persons of similar views. To him it mattered tremendously that those who were themselves to teach should be also able to learn something from their own countrysides and explain this to their pupils. He had himself a perceptive eye for landscapes and was able to retain the impression of them in his mind: increasingly through the years he enjoyed sketching landscapes on his field tours and holidays.

In recent years the position of geomorphology in relation to geography had been a frequent subject of discussion, though in Britain it is usual for geomorphologists to be attached to Departments of Geography. Inevitably there must be some question of the content of geomorphology, if it is to be merely a quest for the recognition of surface features without some genetic explanation of them: naturally, any geomorphologist would wish to elucidate what he sees and to use modern mathematical techniques as one of the means of doing this. If this leads the geomorphologist into geological explanation, then it is surely reasonable that the geomorphologist should use it. Ogilvie was of the opinion that his Honours students should study geology as a contribution to their general geographical education: but while he encouraged geomorphological research, he saw its use in regional geography as something having its own balance with other themes, such as vegetation, agriculture and industry. This was part of his idea of geographical unity, seen especially in his work on regional geography. He told the writer that physical geography, so called, or geomorphology, should always be taught in the Geography Department, as if it was taught in the Geology Department—then the practice in a number of universities —cohesion with the rest of the course was difficult to obtain. On this advice the author, when he moved to Dublin from Edinburgh, acted and became fascinated by the teaching of physical geography and its relation to the other courses. Friendly co-operation with the geologists made demarcation disputes unthinkable: each could help the other.

The regional approach

To some extent, Ogilvie's career followed a continental European pattern, for he began his field work as a physical geographer and became a regional geographer. His essay on central Scotland[1] is a highly informative treatment, based on careful field work. In fact, his first published work on Scotland was a study of Moray Firth,[2] which is a study of the shore forms. At one time, Ogilvie had the idea that he might study oceanography but two days in an open boat were enough. In a reminiscent talk to a student society he exhibited a photograph of a somewhat forlorn figure taken at the time. The essay on central Scotland shows historical perspective,[3] seen especially through its influence on the landscape in the farming evolved from monastic traditions, in the 'Scottish baronial' architecture imported from France, in the trade connections with the Baltic and North Sea ports reflected in the red roofed houses of east coast villages. In the west of central Scotland, however, little advantage was derived from the nearness to the Atlantic until the Clyde was deepened. At the end of the discussion on Clydeside comes a pregnant sentence: 'Nowhere in the world can there be found a better example of the effects of position and of inherent resources in altering completely and rapidly the aspect of a country, directly a new vista is opened in the geographical and economic outlook.'

The essay includes a division of central Scotland into four major physical units, plains of raised beach and alluvium, the 'Lower Lowland' peneplane, the 'Higher Lowland' peneplane: and the plateaus of the uplands. Of the two lowland peneplanes, the Lower lies mainly between 100 ft and 500 ft and the Higher between 500 ft and 750 ft though near to the east coast it falls to 400 ft and around the hills it rises slightly above 1,000 ft. After some consideration of the rivers, glacial deposits and the coasts, it is noted that an

[1] *Great Britain: Essays in Regional Geography*, Cambridge 1928, 409–50. Ogilvie also wrote part of the chapter on southern Scotland, 467–73, 476–9. There are modifications in the second edition.

[2] 'The physiography of the Moray firth coast', *Transactions of the Royal Society of Edinburgh*, 53, part 2 (1921), 1925, 377–404.

[3] His continued interest in historical perspective is shown in 'The Time-Element in geography', *Trans Inst. Br. Geogr.*, 18, 1953, 1–15.

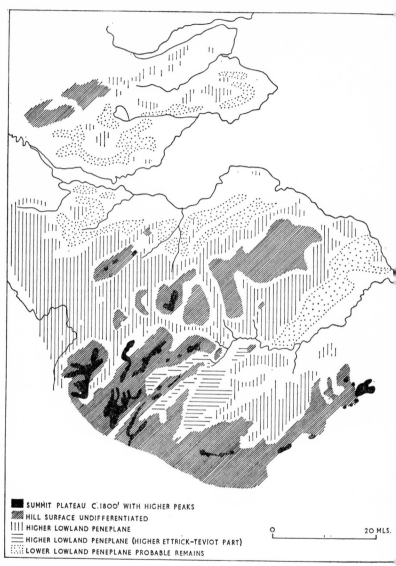

SUMMIT PLATEAU C.1800' WITH HIGHER PEAKS
HILL SURFACE UNDIFFERENTIATED
HIGHER LOWLAND PENEPLANE
HIGHER LOWLAND PENEPLANE (HIGHER ETTRICK-TEVIOT PART)
LOWER LOWLAND PENEPLANE PROBABLE REMAINS

0   20 MLS.

FIG. 20. Surfaces of erosion in southeastern Scotland

area of north Ayrshire has 800 miles of streams in 345 square miles but an area in Lothian has 540 stream miles in 390 square miles. On this Ogilvie comments that

The difference must be the result of a variety of causes. These figures, however, suggest that there is at least some correspondence between rainfall and river density; for the rainfall of the Ayrshire area is between 30 and 60 ins. while that in the Lothian area is between 25 and 40 ins.

The water supply of central Scotland is then considered in relation both to agriculture and the supply of towns and cities. In the sections on vegetation and population Ogilvie follows the evolutionary view characteristic of other writers of the time, and the essay then proceeds to a consideration of agriculture and the 'sub-regions', not here claimed to be *pays* on the French model[1] but introduced as 'the several more or less natural sub-regions into which Central Scotland may be divided'. This section shows some interesting correlations, for example

Strathmore and the Howe are given over to a fairly uniform type of well-managed mixed farming. . . . That basic geographical feature the 'Caledonian trend' is reflected not only in streams and drumlins; but in the diceboard fields, in the roads and even in the set of the farms; these are built of 'Old Red' stone with slate roofs.

And in the description of part of the Tweed basin the same correlation is seen.

The Merse, whose upper limit may be taken as 400 ft, forms one of the largest expanses of drumlins in Britain; and the pattern of streams, roads, plantations, hedges and houses conforms almost exclusively to the trends of the drumlin axes and to the perpendicular directions.[2]

[1] *Great Britain*, op. cit., Cambridge 1928, esp. 432–3.   [2] Ibid., 472.

Fig. 20. In field work, A. G. Ogilvie constantly showed the apparent surfaces of erosion (peneplanes) that characterize much of Scotland, though he was fully aware of the difficulties to be surmounted before they could be adequately explained. With Scottish caution, he presented this map and an earlier one as tentative only: he took the view that the recognition of such surfaces in the field was at any rate a necessary first step in the explanation of the history of the physical landscape. Originally published in Robertson, C. J. (ed.), *Scientific Survey of South-Eastern Scotland*, British Association, Edinburgh meeting, 1951, 14. The earlier map is in *Great Britain: Essays in Regional Geography*, Cambridge 1928, 414.

Later, Ogilvie published two papers[1] on possible land reclamation in Scotland, and also a paper on areas of such gentle slopes that drainage was difficult.[2] This material, with some further observations on the 'lower' and 'higher' lowland peneplanes as they occur in the Tweed valley, were included in an essay published when the British Association meeting was held at Edinburgh in 1951.[3]

Relationships as clear as those (mentioned above) in the Tweed valley and in Strathmore are not to be found everywhere and they may appear to be so obvious, even simple, that they are scarcely worth mentioning. In fact they are not generally noticed until a landscape is subjected to careful analysis, ideally in the field with the map to give a synoptic view. Ogilvie followed this idea of correlation further in a study[4] resulting from a field tour in the summer of 1935, which deals with the southern or Italian face of the Pennine Alps from the summits immediately west of Monte Rosa. The area chosen had a general southern exposure, and ranged from the summits to lowlands having rice cultivation with irrigation. Below the treeline there was a zone of varied coniferous trees, and below this again, an oak-chestnut zone and, in one of the valleys studied, a beech zone also. Many other trees were noted at various altitudes, and the relative prominence of the existing trees had been closely influenced by human choice, for some had been cut for firewood or timber; the treeline had obviously oscillated considerably and had recently risen higher as the local glaciers had been in retreat since the mid-nineteenth century. Most of the upland pastures were used for hay, much of it irrigated: in the eighteenth century, crops were grown whenever possible but the former aspirations for self-sufficiency had given way to a specialized agriculture based on cattle rearing. Along with this there had been a human retreat from the higher levels, for the upper hamlets, formerly occupied throughout the year, had become temporary dwellings for haymakers or had been abandoned. The article is illustrated by a colour map showing the various vegetational types. Within its range, the article is an interesting essay in regional geography, though it does not deal with the population

[1] *Scott. Geogr. Mag.*, 61, 1945, 77–84; 62, 1946, 26–8.

[2] 'Debatable land in Scotland', *Scott. Geogr. Mag.*, 60, 1944, 42–5.

[3] 'South-Eastern Scotland: the region and its parts', in *Scientific Survey of South-Eastern Scotland*, ed. C. J. Robertson, Edinburgh 1951.

[4] 'Natural and cultivated vegetation in the eastern Dora Baltea basin', *Scott. Geogr. Mag.*, 53, 1937, 249–66.

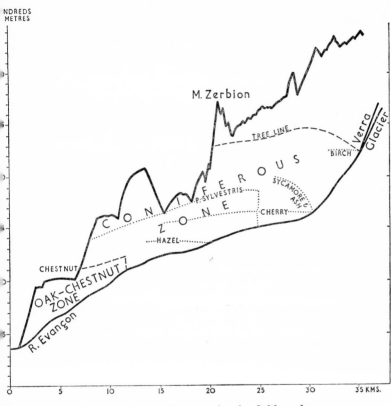

FIG. 21. Vegetation mapping by field work

This is given as an example of work by A. G. Ogilvie, done in the summer of 1935 to clarify the general statements in works on Alpine vegetation rather than to establish new principles. The original article includes a coloured map of vegetation on the 1 : 100,000 scale. Originally published in *Scott. Geogr. Mag.*, 53, 1937, 257.

except in general terms and does not include a detailed analysis of the farming.

Ogilvie was immune from the view that his work was more important than anyone else's and eagerly welcomed new ideas.[1] He was deeply concerned for the welfare of Scotland, and this showed itself not only in work for such bodies as the National Trust and the Royal Scottish Geographical Society, but also in various papers such as one given as the inaugural address when the Edinburgh branch of the Geographical Association was founded in 1937.[2] It had a somewhat provocative title, and dealt with the need for research in the various systematic branches of geography. Geomorphology received little if any support from either the Treasury or official geologists, and as a result little was known of erosion surfaces, of valley density, length of rivers in relation to valleys, the river profiles or glacial phenomena such as the altitude and exposure of corries. Though the Scottish freshwater lakes had been surveyed by Sir John Murray[3] and his colleagues, little was known of silting in tidal estuaries, pollution and other phenomena except for the Forth and the Clyde. On the coasts, regular observation on the effects of storms could be made, not necessarily by professional geographers. The inadequacy of the climatic data—worked out as representing one temperature station to every 930 square miles—was well known, but such work as the observation of the last killing frosts or the incidence of cyclonic spells, could be recorded. The section on vegetation mapping drew a reply[4] pointing out that though a biological survey would be of great value, it was likely to be expensive. But should that deter a government? On population, Ogilvie suggested investigations into the distribution of the rural community between villages and 'disseminated dwellings', and an estimation of the population directly dependent on agriculture. What was the number per square mile in fact supported by farming? The various Scottish dialects could be mapped before

[1] He once told me that he had learned something from every member of the staff who had worked in Edinburgh. What he could possibly have learned from me in 1933–5 I find it difficult to imagine. On the other hand, he was deeply amused by pedants, and once commented, 'Do you notice that So-and-so is *always* giving one information?'

[2] 'Our ignorance of Scotland', *Scott. Geogr. Mag.*, 53, 1937, 387–394.

[3] Their maps of submarine contours were used by the Ordnance Survey.

[4] Fenton, E. W., 'Our ignorance of Scotland—a reply', *Scott. Geogr. Mag.*, 54, 1938, 93–6.

they disappeared and, given the statistics, church adherence[1] could be mapped with interesting results. Enquiries on migration would be facilitated by more material on birthplaces, and the mapping of diseases could be informative—as indeed it has since proved to be through the recent advance in medical geography. On agriculture the work of the Land Utilization Survey organized by L. D. Stamp was acknowledged, but more could be learned of farm sizes and the distribution of farms, and of their assessed worth in the local valuation. Not since 1912, when the second edition of Bartholomew's *Atlas of Scotland* appeared, had there been a detailed map showing the position and character of the mines and factories in Scotland, nor was there any precise knowledge of the degree to which roads and railways were used for the transport of passengers and goods. There is still much to ponder in the last sentence of the article:

My object has been to show that we have a large field for geographical research in our own country, and to try to induce those who are capable of it to attempt to remedy this lamentable lack of exact knowledge and so to advance science, to contribute to social and economic amelioration, and to help to place the geographer's synthetic work upon a sounder basis.

Although some research enquiries suggested in 1937 have since been made, partly by various bodies responsible for planning, in 1952[2] Ogilvie pointed out that when the rivers draining the Lammermuirs flooded in 1948, 'it became quite clear that there is no branch of our administration responsible for study of rivers and supervision of their activities as a whole . . . a deplorable defect in our national and local system of government'. In other respects, however, there were encouraging signs of progress: the British Glaciological Society had found resident observers to report on the local depth of snow; the Third Statistical Account had been launched with the planning of pilot volumes on Fife and East Lothian, the University of Edinburgh had launched a School of Scottish Studies, including

[1] This is known in Ireland through the Census and used by geographers as source material, notably in Jones, E., *A Social Geography of Belfast*, London 1960. Some people think that an enquiry on the Census form would be an infringement of personal liberty, but no objection appears to arise from such an enquiry in hospitals or the older universities, or in Ireland, where liberty is valued.

[2] 'Co-operative research on south-east Scotland: the next step?' *Scott. Geogr. Mag.*, 68, 1952, 22–4.

N

geography; and various planning boards had been established. But the tasks were not completed.

Some wider applications

In the nineteenth century, much of the excitement of geographical study lay in its association with the opening of the world by various new means of communication. Once some of the major facts of world distributions were known, even partially, something more satisfying than travellers' tales must be provided. In 1935 Ogilvie referred[1] to the difference between Europe, where geographers knew their country, had a copious supply of facts and statistics and above all excellent topographical maps, and the rest of the world, much of it inadequately mapped and known mainly through the traverses of explorers. Indeed for large parts of the southern continents and of Asia, much of the available information was derived from accounts of primary exploration, some of the best of which was given by the naturalist travellers of the nineteenth century. Even though expeditions had steadily become more skilled in observation and recording, defects remained. As new areas were occupied and organized, especially in India, Indonesia and Africa, geographers could gather material to make a synthetic regional account in certain 'key' districts perhaps not untypical of larger wholes. Ogilvie wished that young geographers without essential home ties could go to Africa for a year or more on such enterprises, and he was strongly of the opinion that they could be usefully employed not only on the obvious task of organizing topographical surveys but also of studying the human geography: he lived long enough to see at least a tentative beginning in this enterprise.

Meanwhile he recognized that there were many people in close contact with the Africans who had a soundly-based knowledge of the local environmental conditions, particularly the District Officers and missionaries. With the aid of a Research Committee of the British Association, a questionnaire stressing nineteen points and a pamphlet explaining the aims was drawn up: this included reprints of two essays on the relation of African tribes to their environment, by Père L. Martrou on the Fang and R. V. Sayce on the Basutos. By the

---

[1] 'Co-operative research in geography, with an African example', *Scott. Geogr. Mag.*, 50, 1934, 353–78. The idea was mooted earlier, 'Africa as a field for geographical research', *Geogr. Teach.*, 13, 1926, 462–7.

early 1930s the most comprehensive response had come from Northern Rhodesia in the form of thirty reports from District Officers. At that time the physical geography of Northern Rhodesia was less well known than that of Katanga, in the then Belgian Congo, on which some excellent work had been done. Some extremely interesting information was acquired from the Rhodesian reports on the agricultural activity of the people, their use of savana timber, their houses and their migrations to earn money, in many cases over long distances. It was also possible to estimate the population density, the stock of cattle and its use and the fluctuating distribution of the tsetse fly: there was even a map of the main food staples consumed. Essentially the enquiry provided new geographical material on an area imperfectly known but it also gave some understanding of the material conditions of native life, in fact highly varied from one area to another, not only through environmental differences but also from those of outlook and tradition. Like many other enterprises, this enquiry did not survive the war, and comparable work must now come largely from the universities established in Africa, which obviously have opportunities almost beyond imagination for profitable work.

In his last years, Ogilvie was working on a book dealing with Europe,[1] but he died before it was completed and the manuscript was therefore prepared for publication by his colleague and literary executor, Dr C. J. Robertson. Throughout his teaching life, Ogilvie had been concerned largely with Europe at various levels from the relatively simple work with the first year (very) Ordinary students[2] to advanced work with final Honours students in their fourth year, devoted primarily to regional courses. Early in life he acquired a strong command of French and German to which he later added Italian. The posthumous book includes Europe to its traditional eastern limit in the Ural mountains and the Caspian depression, though as the author admits this is no longer realistic in view of the political and economic organization of the U.S.S.R., really to be thought of as a continental area in its own right. Also included as the Borderlands are the Caucasian region, the whole of Turkey and Iraq, with a strip of Iran, Syria, Lebanon, Jordan, Israel and Egypt,

---

[1] *Europe and its Borderlands*, Edinburgh and London 1957.
[2] The custom that the senior geographer should be responsible for lecturing to first-year students was maintained in the Geography Department at Edinburgh.

which together form the 'bridge-lands of Asia'. The settled coastal area of Africa in Libya, Tunisia, Algeria and Morocco are included, so that in effect the southern frame of 'Europe and its borderlands' is the desert in Africa and Arabia, and the northern frame is oceanic, though interrupted by the islands surrounding the Barents Sea and by Iceland.

The purpose of the work is to study 'the natural phenomena which, by their action and reaction, compose the physical environment of the western peoples', one-fifth of the human race, one-quarter if the western parts of the U.S.S.R. are included. It is easy to recognize that the great concentrations of industry in western Europe, and for that matter in the U.S.S.R., are among the greatest population clusters in the world, apparently destined to grow and spread more and more in the future. But in Ogilvie's view

most of the characteristics of the Western World existed long before these lands were densely peopled by present standards; they grew from the good ideas of a few men . . . the locations of the Greek city-states, of Israel, of Rome, are facts of geography that bear upon the subject at least as much as the position and content of the northern coal-basins.

No continent offers more problems to a writer than Europe, as its variety of scene, both in its physical features and through the imprint of human activity, is virtually unlimited. The present author, leading a party of overseas geographers through the north of England during the International Geographical Congress in 1964 found amazement at the variety of scene, not to mention the strong historical imprint, in an area which some textbooks had represented as devoted to cotton on the west and to wool on the east side of the Pennines. And there lies the difficulty, for generalization can conceal as much as it reveals.

Nevertheless, no book on Europe can be written without some generalization. *Europe and its Borderlands* gives a synoptic view, and two-thirds of it is given to chapters that can be grouped under the general headings of physical geography including climate and vegetation, human activity, states and frontiers, economic aspects with transport and population. There follows a regional treatment as such, though it should be noted that many of the topics in the first two-thirds of the book are treated regionally, at least in part. It is well illustrated and some maps such as that showing the density of woodland in three categories—almost continuous, woods predominant, considerable woods—with the fourth category (few woods) unnamed, will mean a lot to those who have travelled sufficiently to realize how varied the imprint of woodland is from one part of Europe to another. Within Great Britain only parts of the Scottish Highlands come inside any of the three major categories, and there only the third, considerable woods, group. Throughout the book, the wide reading and vast experience of the author is apparent. However detailed geographical work must become, it can be argued that to possess a general conspectus of the geography of Europe and its fringes can still be valuable and—within its range—satisfying.

To an Englishman, Ogilvie seemed very much a Scotsman and his strong national consciousness was seen in his devoted work for those Scottish institutions, such as the National Trust, which were preservers of the country's scenery, traditions and dignity. For the Scottish Geographical Society he had the highest aspirations, and deplored the periodical outbursts of a 'popular' policy found in its council. He encouraged the editorial policy of including articles on a wide range of themes by writers from outside Scotland in harmony with Dr Marion Newbigin, Editor to her death in 1934 and her successors, Miss H. G. Wanklyn (Mrs Steers) and Miss L. R. Latham: naturally he encouraged all possible research on Scotland. It was a source of disappointment to him that few of the capable, strong young students who graduated in the Honours school found opportunities for work in the Dominions and colonies, but after the 1939–45 war to his satisfaction geographers found a wide range of posts. He gave devoted service to such enterprises as the International Club which worked for the large contingent of overseas students in Edinburgh. Breadth of mind he appreciated in everyone, and partly for this reason he was glad to welcome a number of English students to his classes as they might modify the provincial outlook of some Scottish students, especially those from small towns and dreary industrial areas. To many, he was a confidential friend as safe as a father confessor. Quiet and circumspect in manner, he was generous with his time to others and like many more academics regretted that the time for writing was restricted. He gave loyal service to the International Geographical Union, the Institute of British Geographers, the Geographical Association and other professional bodies and enjoyed the meetings and social contacts they provided. Though physically active, he was rarely in full health and his infirmities increased after the death of his devoted wife in 1952. He died suddenly at the end of a full day's work on 10 February 1954.[1]

[1] Obituaries and appreciations include those in *Geogr. J.*, 120, 1954, 258–9; *Geogrl Rev.*, 44, 1954, 442–4; *Scott. Geogr. Mag.*, 70, 1954, 1–5; *Trans Inst. Br. Geogr.*, 20, 1954, viii–ix; Miller, R. and Watson, J. W. (eds.), *Geographical Essays in Memory of Alan Grant Ogilvie*, London, 1959, xi–xvi, 1–6.

CHAPTER 8

The Varied Views of Geographers

THE intention of this work has not been to write a series of panegyrics on a limited number of geographers but rather to choose a few workers and show what they did in their own day and generation. Every one of them was to some extent a child of his time though each reacted to circumstances according to his own temperament and character: contrary to a view widely held by those outside universities, there is no standard academic type but a wide range of individuality, not to say eccentricity in some cases. It would be highly undesirable to cast academics into the type of mould that some occupations appear to demand, or at least favour, though the one thing that all academics need is the sincerity of purpose which all those treated in this book obviously possessed. In various ways they were interested in presenting their views to the rest of the world and in doing so they naturally revealed themselves. With the exception of Francis Galton, all were geographers by profession if not by initial training and possibly some of them were more effective as geographers because they had in their student days studied other subjects. The battle still wages in academic circles on the desirability of intense specialization at an early age and there must be many geographers who, like the present author, look back to the stimulus given by teachers of other subjects, notably in his case those of history. All those considered here also had a knowledge of at least one foreign language, and that suggests that such knowledge can open many fields of interesting study even though English-speaking students now have a vast range of literature available, either from American geographers or from continental Europeans, such as Dutchmen and Scandinavians, who write largely in English. To a great extent the geographers of an earlier time drew their inspiration from the workers in the German universities before 1914 though after 1918 the French geographers were more thoroughly studied: in our own day, many turn constantly to the work of the Americans as they experiment with new techniques, with varying success. Meanwhile, the number of geographers in the world has vastly increased and the academic scene is no longer dominated by a few great figures, though there are some who would like it to be.

The work of seven geographers

Of the geographers treated here, Francis Galton alone was an amateur in the sense that he was endowed with vast wealth and worked when he felt like it. He was one of many Victorian travellers who blazed new trails, and the attention given to his journeys through Africa show the hunger for vicarious adventure that the British public possessed at the time, due in part to a desire for new markets, in part to the considerable missionary enthusiasm then prevalent and also to imperialist fervour combined with a wish to spread civilization into the darkest corners of the earth. But there was also the appeal of courageous enterprise and natural curiosity about places where no white man had penetrated before: it is strange that Galton never went again to Africa or any similar region little-known at the time, but he appears to have enjoyed the leisure and comfort that he could command and it apparently never occurred to him to use his long sojourns in continental Europe for any geographical purpose. The idea that the geographer can do more work of significance in remote places than in his home environment still survives quite widely, even in universities, probably because new observations can be made more readily in areas little traversed than in those on which many people have worked already. What nineteenth-century travellers provided was raw material, and Galton was one of a group of people who through the Royal Geographical Society in London encouraged travellers to bring back the type of observations that could provide basic material for a geographical synthesis, along with exploratory maps that could be preliminary to more adequate surveys of a later time. Through such enterprise the work of nineteenth century explorers was much more useful than it would otherwise have been: unfortunately the geographical journals of the nineteenth century are packed with trivial information on journeys with occasional observations of real value. Nevertheless, the conquest of desert, mountain and ice cap was an entirely laudable enterprise and one that added greatly to geographical knowledge.

Galton's meteorological atlas[1] was a remarkable achievement, largely because his self-imposed task of mapping the weather day by

[1] See pp. 30–4.

day led to the discovery of the anticyclone: had Galton continued this line of enquiry he might also have discovered the mechanism of depressions as his work on such pressure systems came within sight of the facts. The atlas is beautifully produced, and there may not now be many copies available. Though Galton maintained a lifelong interest in meteorology, he turned his enquiring mind elsewhere and became fascinated by race, heredity and eugenics, all of which were associated with the outburst of interest in man as himself a scientific product, now much lower than the angels, which was stimulated though not initiated by the work of his cousin Charles Darwin with whom, for part of his life, he was in friendly association. Less constantly ill than Charles Darwin but prone to enjoy ill-health, Galton and his cousin could enjoy a chat about their symptoms. In the work that could be grouped for convenience as anthropological, Galton produced a number of maps, including the one reproduced in this book.[1] He also did a pioneer isochrone map and permanently maintained his interest in geographical education, but his work in geography was only one aspect, and in some ways a minor one, of a life spent as a cultivated man of science unimpeded by the need to earn a living.

Vidal de la Blache stands out as one of the greatest figures of what might be called the modern geographical renaissance. In his initial university education, he studied geography with history, but the teaching in the former apparently was so pedestrian that it might have killed his interest in the subject. His post-graduate residence in Athens and Rome gave him an abiding interest in the history and the geography of the eastern Mediterranean which was linked with a knowledge of the ancient classics. A good linguist, he was a writer of distinction and his introductory volume to the great Lavisse history of France[2] is generally regarded as a contribution to literature as well as to geography. In his lifetime two major tragedies came to France, the Franco-Prussian war of 1870 and the war of 1914–18: following the gradual move to Paris of many distinguished French academics, he lived first at Nancy where, in an environment very different from his native Mediterranean, he saw the effects of the loss of Alsace-Lorraine. His main work on this area appeared long afterwards, in 1917, and was in effect an argument on geographical and historical grounds for the return of Alsace-Lorraine to France.

[1] P. 37.   [2] Pp. 54–6.

He died before the end of the 1914–18 war, in which personal tragedies came his way. As a writer on France Vidal de la Blache is known widely and his scheme of regionalization based on the local unit, or *pays*, is renowned. Criticisms made of it at various times appear to be based on a scanty knowledge of the *Tableau de la géographie de la France*: in fact, it was never claimed that the *pays* were uniform in structure or land use for some were varied by surface deposits on the underlying rock and, though regarded as units, were of heterogeneous rather than of homogeneous character. The basis of regionalization lay in the historic realization of various local groups that their particular territory was different from that of their neighbours: in short it was a regionalization based on physical features, differences of agriculture, and the knowledge of the people themselves.

But though the work on France stands out as classical, much more was done by Vidal de la Blache. In particular, the *Atlas Vidal-Lablache*, first published in 1894, has been studied by one generation after another and is still on sale in a revised form: comparisons with other historical atlases only serve to deepen the admiration it arouses. For successive periods, the distribution of states, towns, commerce and settlement is shown and a brief but effective commentary accompanies each map. In retrospect the historical section may seem more remarkable than the geographical section, though that summarizes much of the knowledge of the time. It may in part have been this atlas that inspired some geographers to attempt a study of the geographical influence on the whole of human history, at least in lecture courses. Is a geographer wise to attempt so much? Some years ago at a meeting H. J. Fleure said that man had been on the earth for half-a-million years (now commonly believed to be a million years) and the geographer must study his distribution and works over all that time. But shortly afterwards, as on other occasions, C. B. Fawcett said that the last hundred years were of greater significance than all that had gone before and the task of the geographer was with the nearer time. In fact some landscapes show the clear imprint of a far longer time than a century and Vidal de la Blache doubtless saw the imprint of the Roman and even, in places, the Greek period in the Mediterranean area of France and especially in Italy and the peninsulas and islands farther east towards the Levant and Palestine. Nor does it seem reasonable to study a town like

Chester without some enquiry into its Roman occupation, especially as the existing street pattern bears some resemblance to that of the Roman period, and their walls are still visible in places beneath the medieval walls that form a complete circuit of what is now the central part of the city. And in Dublin it requires little imagination to reconstruct the Norse town on a slight hill round an old but rebuilt Cathedral, to see the Liffey as unconfined and spreading over marshes now banked up and drained, and to visualize the wharf on the river below the Cathedral and the sheltered inlet further downstream where boats could be moored. In short, it has taken nearly 1,900 years for Chester to evolve, and more than 1,000 years for Dublin to reach its present form. The earlier prehistoric periods now belong to the archaeologists, many of whom have evolved a fine technique of environmental reconstruction. The trouble with some geographers is that they possess little historical perspective, though some fine work has been done in historical geography conceived as the regional geography of the past, in effect a kind of environmental reconstruction.

Other work by Vidal de la Blache covered a wide range of human distributions and activities, but his main work on human geography was published posthumously and was not complete at the time of his death. It was a view of the entire world, based largely on wide and dispersed reading, and expressive of his view that all phenomena on the earth's surface are interrelated with man drawing his sustenance from the world and changing it from one age to another. The varied results that come from man's use of the environment were clearly a source of fascination to Vidal de la Blache and his demonstration of their variety has led to his association with the 'possibilist' theory of the relation between man and environment. To some extent this was due to his inherently historical approach, for he shows how peoples have had times of prosperity followed by others of failure and disruption. He had much of the logic and subtlety of approach that one associates with French writers and in all his writing he shows an awareness of the fundamental physical environment: as one reads and, perhaps wisely, re-reads his work on human geography it appears to spring from a vast knowledge of local, regional geography, taking examples from one area after another to illustrate a geographical principle as an historian might take one event after another to illustrate an historical principle. The continued success of the

Annales de Géographie, founded in 1891 as a serious academic journal to supplement the then numerous but less academic journals of various geographical societies in France, is a continuing tribute to Vidal de la Blache and his colleagues who founded it: he did not live to see even the publication of the first volume in the *Géographie Universelle*, the series of regional texts covering the whole world, which he planned with others who carried on the work after his death.

Jovan Cvijić was a geographer whose career was deeply influenced by the needs of his country though his initial work was on the richly varied physical landscapes of the Balkans. Anyone visiting a limestone area will be curious to know how such land forms could have evolved and this was the problem on which Cvijić did some of his most famous work. He also worked on the problem of the glaciation of the Balkan peninsula and successfully demolished the existing view that it was unglaciated. But the work that eventually made him an internationally known figure developed from his regional study of the Balkan peninsula, no longer Turkish-dominated and favourably placed for independence if Austria-Hungary were defeated in the 1914–18 war. Cvijić had for many years been fascinated by the human problems of the peninsula, with its highly mixed populations scattered through migrations and wars, preserving many traditional ways, of mixed allegiance in religion, with transhumance in the mountain areas. The investigations began during long tours of the country on foot, and covered a wide range of enquiries including much that would now be classified as sociology, physical and social anthropology. When the new state of Yugoslavia was defined after the 1914–18 war, maps showing the distribution of languages, ethnic groups and religious allegiance, were produced in evidence, and the work of Cvijić and those who had gathered round him became of supreme significance. His personal contribution was the more effective because it was presented academically rather than with the political passion then so prevalent. *La péninsule balkanique* ranks as one of the great regional geographies of its time and Cvijić is a fine example of a worker who used the material of his homeland with perspicacity and indeed enthusiasm.

Enthusiasm undoubtedly marked Ellsworth Huntington as a worker, and to a European he seems to be almost the epitome of the large and powerful American academic type who may be seen at international congresses. His output of work was enormous, partly because he was freed from many of the routine duties of most university teachers. His aim was nothing less than the explanation of all human activity and, inspired by his early explorations in semi-arid areas with fluctuating water supplies, he found the explanation he needed in climate. Not for him was the cautious view with which Marc Bloch ends his book 'the causes cannot be assumed. They are to be looked for . . .': he had found the causes of human action or the lack of it, even in such trivial matters as the borrowing of library books or the more serious aspects of delinquency and immoral conduct. He was closely acquainted with the racial theories of his time, but saw physical development largely in climatic terms. And yet, as one reads his works, they induce a certain admiration: he knew his own mind and was never afraid of the apparently unanswerable questions, such as the reasons for the decay of the Maya or Aztec civilizations. It may be true that in the time of Christ Palestine was better watered than now, and that the reasons for the decay of Mesopotamian, Egyptian, Greek and Roman civilizations were partly climatic, but many other possible causes have been adduced. Certainly the relations of the peoples of the interior steppelands and deserts of the Old World with the people of the settled agricultural margins open a fascinating study, and Ellsworth Huntington's treatment of such themes gave him success on a scale rarely accorded to geographers: but there are those like the present author who could echo the remark of an examiner to a research student, 'It's all very interesting but I simply can't believe it.' Or at least, not all of it.

If Ellsworth Huntington stands out as a man possessing great assurance with a mind sweeping across the world and through human history like a whirlwind, Sten de Geer stands out as a man of caution, patience and practical good sense. Like many other geographers he worked first as a geomorphologist and then turned to broader studies, and notably to the distribution of population in Sweden on which he produced a remarkable atlas in 1919. The cartographical presentation by dots for every 100 persons in the countryside inevitably involved some generalization in a country having many isolated farmsteads in areas mainly covered with forest, but within such unavoidable limitations Sten de Geer's work gave a far more accurate view of the distribution of population than any other method:

anyone who has tried it elsewhere, as the present author has done for Ireland, will have found that it produces distribution patterns of considerable interest. And it forms a basis for a regional treatment of a country that can be useful and interesting, as Sten de Geer demonstrated in later works. Even more, if done for one time, such as 1917, it can also be done for some later time by when, as is certainly the case in Sweden, the distribution of the rural population may have changed markedly, probably by heavy loss in the farmed areas of some prosperity, the withdrawal of people from the poorest areas and the increase of villages and hamlets. The atlas must eventually rank as an important historical document for future workers.

But Sten de Geer went further, and shared the interest in racial characteristics prevalent in the 1920s, and with this he considered the possible political implications of differences of racial origin (real or imagined), religion and national character. He was able to view these in a far more detached manner than Jovan Cvijić, for problems of defining Scandinavian countries had ceased to be a living issue in Europe, partly owing to the sensible attitude of the inhabitants. Another enterprise which shows the broad sweep of his mind was a study of the American manufacturing belt, done during a sojourn in the United States and crystallizing a large amount of detailed enquiry and mapping into a synthetic statement. This work, as a contemporary picture, is of potential—even actual—historical significance. Within a short life of only forty-seven years, of which the last few were given largely to work for the advancement of geography at Göteborg rather than to research, Sten de Geer spread his work from minute studies of rivers to European political geography and the economic geography of the United States. Which part of his work has the most permanent value? To the present author it would seem that the atlas of population stands out as his supreme achievement, as it enshrined a great deal of field work and mapping, related a major human distribution to physical features and gave a contemporary picture of interest to students of a later time.

The two British geographers discussed in this book were both men of great sincerity, both widely travelled, and both deeply attached to the areas in which they spent the greater part—in Roxby's case almost the whole—of their working lives. But neither was provincial. Both of them had interests in distant areas, and both regarded the improvement of international relations as the concern of every citizen: indeed both gave a substantial part of their leisure time to such enterprises as social work for overseas students. But their experience was vastly different. Circumstances led Roxby to China at a time when it was little known geographically, and a short time afterwards Ogilvie, described to the author by a contemporary as a young, bright officer, served throughout the 1914–18 war. Both favoured a regional treatment of geography, and that of Roxby was seen on a broad scale in his work on China and on a more detailed scale in the *pays* treatment of East Anglia. Ogilvie, far more deeply versed in geomorphology, used it as the basis for regional study, conspicuously in central and southeast Scotland though his other works, on the central Andes and Europe, also have a strongly regional character. Each was aware how little was really known of the world, though Ogilvie was the more prone to say so and even—courageously —to point out that Scotland was only partially studied. Both maintained an interest in the historical aspects of geography expressed in Roxby's demonstration of the gradual penetration of what became China by a distinctive type of life with a socio-religious ethos, and in Ogilvie's presidential address to the Institute of British Geographers, which noted the survival of many past boundaries, roads and other features into the landscape of the present.

A time for experiment

Seven geographers with seven mops could not possibly sweep away all the cobwebs of geographical ignorance that the world possesses. Frankly, we know so little and we take so much for granted. No longer are we sure that there is a simple correlation between climate and crops, still less between climate and man, and even the world classification of climates, so long the foundation of world regional schemes, seems dubious. And climate is itself an ephemeral phenomenon, not something that can be classified into a series of regions that were once regarded as virtually permanent. In countries such as Norway, the climate is obviously oscillating to a remarkable degree from year to year and the general decay of glaciers and ice sheets in the northern hemisphere suggests that a period of decreasing cold is now in being. But no one really knows to what extent this is due to changes in ocean currents, which may also affect the distribution of fish, notoriously subject to booms and slumps, though periods of stringency may be due to the greed of certain nations which, by a kind of robber economy, prevent the natural regeneration of fish supplies. That climatic data is inadequate no one would dispute and with more data the initial synthesis becomes unsatisfying and much has to be done before a new one emerges.

Almost with surprise, many people began to realize in Britain during and after the 1939–45 war how little was known of our towns, despite the historical-evolutionary studies to which many had been subjected. No doubt many historians would make the same comment about the history of towns, with a few conspicuous exceptions. Work that comparatively few geographers ever thought of doing had to be done quickly as a basis for the future planning of towns and in this, as in much more of their activities, planners became applied geographers. In Paris, a young geographer wrote a geography of a railway station,[1] and showed it to be a deeply human phenomenon, its location in a particular place dependent partly on physical geography and partly on the development of Paris, its goods and passenger traffic interpenetrating the life of the city with that of France and even of Europe. The relation of towns to the tributary areas they serve, and the transport by which this is achieved, all became absorbing features of study, and even the movement of pedestrians along particular streets in towns shows a correlation with site values and rents. The presence in a town of a major general store gives it a high status as a shopping centre, and there is at least one borough in England where the town clerk deplores the lack of a Marks and Spencers, as without it there is little prospect of becoming a county borough. The distribution of professional football grounds, athletic stadia, large cinemas, theatres, or even halls for bingo, ten-pin bowling or whatever may be the latest fatuity, are also significant and worthy of study.

If, as is commonly argued,[2] man made the earth, he undoubtedly made the towns and is now re-making them as fast as economic circumstances will permit. Those young enough to see the changes of the next fifty years will undoubtedly witness a transformation of vast scope, though whether the prophecies of Peter Hall[3] will be realized is open to question: the present writer shares the view[4] of

---

[1] Clozier, R., *La Gare du Nord*, Paris 1941.   [2] Pp. 59–60.
[3] Hall, P., *London 2000*, London 1963.
[4] L. D. Stamp, born 1898, died in the summer of 1966.

L. D. Stamp that if Hall is right, he is glad that he will not be here in A.D. 2000. In Sweden, houses are built to last no longer than seventy years at the most: in fact many commercial buildings have there, as in Britain, an even shorter life (and in America shorter still). The result is an urban environment ephemeral in character, constantly subject to change. But not altogether. It is hard to imagine Paris without the boulevards that were the result of an inspired nineteenth century plan, and no people of such taste as the French could bear to remove many of the historic buildings of the central area. And in such cities the natural and semi-natural landscape features such as the Seine and the Thames, the heights such as Montmartre, Montparnasse, Hampstead or Highgate, must be used intelligently in making the new cities, just as the fine sites of Glasgow, Edinburgh or even Birmingham and Sheffield must be used in the future more intelligently than they have been in the past.

At all times, geography is a study of distributions. Distance is not something conquered for all time by rapid means of conveyance, but itself an expression of the interrelationship of distributions, of one town with another, of one farm to the town its people patronize, of the way taken by a migrant from a village seeking work in a town or of the pleasure-seeker going from his home to a town centre for entertainment. None of us lives in isolation, given the normal social instincts: all of us go forth to work, to play, to be edified or amused, or merely to go somewhere. The new M6 motorway in Britain has brought the Lake District within a day's reach of the motorists of Birmingham and the Black Country so that its resorts are more crowded than ever before, and as one area of natural beauty after another becomes popular and accessible, people go even further in this holiday conquest of the world. Even so, where are the Lake District visitors? Those who come for the day from the West Midlands may not penetrate far though they will add to the congestion of Windermere, Ambleside and Grasmere. But even if the more accessible peaks are crowded, many more are visited by comparatively few people even at the height of the holiday season, as the author knows from experience. A survey of the actual distribution of the visitors, organized with a team of workers, might produce interesting results. All is changing, fast, and the geographer serving this present age is confronted with a different range of problems from those of his predecessors. But he is equipped with more techniques, such as aerial photographs, machines for statistical analysis, improved and increasingly informative maps and much more.

The idea of relationship has long been prevalent among geographers but in recent years there has been a swing away from regional work to a cultivation of systematic branches of the subject, and these in time have become increasingly specialized, so that what was once called 'human' geography may now be represented by a dozen or twenty specialisms. Yet the essential quest still remains: why is some observable feature, such as a farm or factory, where it is? Each is a product of human enterprise in which certain difficulties have been met, certain opportunities met with varying degrees of success. Its circumstances may change and its activity expand or contract to extinction. No study can be valid unless it takes account of the individual farm or factory, each field and moor. On a coastline exact mapping of raised beaches and past shorelines may be necessary, and once that is done it may appear that the deposits on the raised beach give a soil worked by the farmers into a richness that gives them great value. On some of the Norwegian coasts, virtually all the agricultural settlement appears to be on alluvial fans made by the streams that gash the hillsides or on the strandflat, recognizable as a restricted patch of lowland though its origin is still controversial. Here is a relationship, visible and clear on the ground, mappable through careful and time-consuming field work, demanding the cooperation of specialists for a full explanation, who can in the end give materials for a new and more satisfying regional synthesis than that of the past. For to condemn or ignore the workers of the past may be to thwart the workers of the future.

Index